THE Life & Times of the Neutron

universal energy

By

Keith Dixon-Roche

THE PROTON-ELECTRON PAIR

Keith Dixon-Roche © 2017 to 2026

THE Life & Times of the Neutron

Universal energy

Published by CalQlata

info@CalQlata.com

First published August 2019
Final publication April 2026

Contents

Preface

Neutrons are something of anathema; we know they exist, but unlike electrons and protons, they aren't lying (or flying) about us. And yet, we don't have a clue what they are, why they exist or where they come from …

… or do we?

Despite their apparent insignificance, I have always suspected that they probably play a far more important role in the workings of our universe than simply providing electrical isolation between adjacent protons. And – as it turns out - this happens to be the case. For example; they were responsible for the last 'Big-Bang', and thereby all universal energy today, and in fact, throughout this universal period.

I hit upon the true nature of the neutron when - following my discovery of the proton-electron pair - I also discovered the neutronic radius, at which instant the magnetic field force generated by the pair exceeds the electron's centrifugal force (chapter 6.5), which occurs when the electron is orbiting its proton partner at the 'speed of light' (c), and is the purpose of Henri Poincaré's famous - but misunderstood - formula; $E=mc^2$; it applies to potential energy, not kinetic energy. In circular orbits the potential energy between a force-centre and its satellite is twice the satellite's kinetic energy;

$$PE = 2.KE = 2 . \tfrac{1}{2}.m.v^2 = m.v^2 \qquad \text{(chapter 2.5.7)}$$

$$\text{and at the neutronic radius (R} \square \text{); } PE = m.c^2 \qquad \text{(chapter 6.2)}$$

This occurs when an atom's [two] innermost orbiting electrons unite (at the neutronic temperature) with their proton partners, storing their energies (kinetic + potential + spin) at the time of their union.

When a neutron - that is created and can only exist within an atom - is ejected, it will revert to its constituent parts a proton and an electron, and the atom in which it resided changes its atomic identity (Z or ψ). This is the cause of radioactive decay in; atoms (natural half-life), nuclear reactors (controlled) and atomic bombs (uncontrolled).

Having discovered how they are created; it wasn't difficult to realise where it occurs (bright stars) and how to harness their energy (heat).

To conclude; you don't need to extract this energy from the critical mass of radioactive matter; it's everywhere in the universe. It exists in all the matter around us. It needs no mining or processing, it's in rock, waste material, soil, garbage, plastic, etc. There is enough energy in less than a decimetre of the earth's dry surface matter to last mankind until the next 'Big-Bang', or for as long as we survive. And it is free, clean and safe.

This book is the conclusion of my atomic studies, which centres on "the life and times of the neutron", and for those of you that missed it, is a pun on its responsibility for atomic half-life.

The original purpose of this book was to raise awareness to the existence of an infinite (well nearly) source of clean, safe, free energy. In fact, the only such source in the universe, and it is everywhere. I have since concluded, however, that this was a pointless quest. Such a source of energy is not in the interests of the world's "leaders". Industrialists and politicians are making far too much money for themselves out of swapping one pollution for another and controlling all aspects of energy to want it made universally available. But, just in case there is someone out there …

Keith Dixon-Roche 2026

1   Introduction

I have always believed that somewhere there lies an undiscovered energy source that is cheap, clean, safe and "infinitely" abundant. My recent work on the atom provided the answer; the neutron.

A neutron is a proton-electron pair that was united at high temperature, retaining the energy the pair were generating at the time of their union, and which it releases when ejected from the atom.

The first thing to understand about neutrons is that they cannot exist outside an atom. Unlike the proton and the electron, you won't find any lying around in your back yard.

Neutronic energy is released within atomic bombs and nuclear reactors, both of which exploit the critical mass of radioactive matter to release it. The difference between the two processes is:

An atomic bomb releases this energy by achieving critical mass almost instantaneously; resulting in an uncontrolled explosion of the matter, breaking it apart and leaving a great deal of unexploded radioactive matter distributed over a large area.

A nuclear reactor releases the same energy but in a (relatively) controlled manner by achieving almost critical mass slowly; resulting in the release of electro-magnetic energy; heat.

The problem with both these processes is that they use a pile-hammer to crack a nut. They are extremely difficult to control and both rely on the use of the critical mass of radioactive matter; making them dangerous and impractical. Moreover, their fallout is radioactive.

If on the other hand we understood what neutrons are; how they're made, and why & when they release their energy, we would be able to use them to our advantage, safely and practically. Moreover, we can use fission to extract their energy from any matter, just as occurs within stars.

Note, fusion is not a source of energy; it requires the input of energy to work.

The primary benefit of this form of energy is that its fallout is hydrogen and/or helium (proton-electron pairs). It is both free and clean, and there is an inexhaustible supply here on Earth.

We can acquire all the energy we need from neutrons within the earth's crust in any element or combination of elements available (rock or even soil); they do not need to be radioactive because we need not rely on critical mass to release it.

THE PROTON-ELECTRON PAIR

Whilst batteries, solar cells, wind turbine generators and power stations are, dirty, expensive, unreliable, wasteful and inefficient; neutron energy is clean, safe, reliable, eternal and free (it can be extracted from your garden). Moreover, it is massively efficient (>231,000,000%), once initiated it fuels itself, and you can switch it off safely, as and when required.

There is enough neutron energy in less than one decimetre of the earth's [surface] crust to supply our energy needs for more than an entire universal period. For example;

1kg of iron, which holds 14.5253 GW.hrs of neutron [heat] energy, is theoretically sufficient to run a car for 1106.75 years, an average UK household for 4418.67558 years or a Jumbo Jet for 90 hours.

Even if only 1% of this energy can be harnessed, it substantially outperforms all existing energy sources, but more importantly it is safe, free and totally clean; it needs no mining or processing.

Given that the human race is unlikely to survive beyond a further million or so years; we have an unlimited source of energy available to us; and everybody can access it safely for free.

Because neutron energy can be conclusively proven (chapter 6.6.1, the neutron as described in this publication must be genuine. Which also proves that E=mc2 applies to neutronic orbits and 'Rn' proves that these orbits must be circular.

Therefore, Newton's and Coulomb's orbital model must be correct and Einstein and Bohr must both have been incorrect.

To conclude: the energy available in neutrons is genuine and exploitable.

1.1   What can we do with this knowledge?

Given what we now know about this energy;
1) How it is created (spin-friction)
2) Where it is created (bright stars and planets)
3) How it is transmitted (electro-magnetic energy)
4) Where it is stored (neutrons)

What can we do with it?
1) Manufacture energy cells of any capacity
2) Recycle any and all forms of waste, including nuclear
3) Eliminate; batteries, solar panels, wind & wave turbine generators, national power-stations, power transmission lines, fossil fuel recovery systems, refuelling stations, and in fact, more than 99% of all today's energy generation systems.
4) Eliminate the risk of fire in transportation accidents
5) Reduce the manufacture of heavy metals by more than 99%
6) Power your house for life from the matter in your garden
7) Virtually eliminate mining and processing
8) Eliminate dangerous by-products (neutron energy $\rightarrow$ hydrogen/helium)

2   Energy

Energy was a concept unknown to Isaac Newton, so he used force to describe energy transfer; which is the manifestation of energy transferred between two or more bodies separated by physical distance.

Our universe comprises electrical and magnetic energy and nothing else; there is no such thing as mass or gravity.

Mass is a term of convenience for atomic particles - and collections thereof - that was derived for an unknown property, and gravity is the magnetic attraction between them.

Mass actually refers to packets of magnetic charge. Max Planck referred to these packets as Quanta, which is the term that will be used here to collectively describe the only two that are required to make the universe work; the proton and the electron.

Mass is therefore, the resistance to movement of Quanta that is under the influence of universal magnetism (gravity). Force is the effort required to overcome this resistance, and energy is the amount of effort applied.

Apart from the non-polar magnetic charge present in all Quanta, electricity and magnetism are polar; negative or positive. Opposite poles attract and similar poles repel.

Because Quanta has opposite electrical polarity, when encountering other Quanta, their electro-magnetic energies will always be opposite or identical according to nature's requirements; i.e. polarity conflict is impossible.

The [orbital] kinetic and potential energy in all universal matter collectively equals the [neutron] energy released during a 'Big-Bang', and remains constant throughout the subsequent universal period.

All universal heat energy is generated by spin-friction within celestial satellites, stored in neutrons and released during subsequent 'Big-Bangs'.

All bright celestial bodies are satellites with sufficient sub-satellite mass to generate fission in their core atoms through planetary spin. Celestial bodies with no force-centre, or satellites with no sub-satellites, generate little or no EME.

The entire universe comprises a fixed, unchanging quantity of energy ($\approx$7.4E+60 Joules), it always has done and always will. It was originally contained within $\approx$3% of the neutrons in the ultimate-body, released during the latest 'Big-Bang'; it remains unchanged today. [first law of thermodynamics]

THE PROTON-ELECTRON PAIR

Environmental EME is trapped by orbiting electrons, which they convert into kinetic energy, and instantaneously - together with their proton partners - generate EME of their own (of exactly the same magnitude), and which they radiate back into the environment. This is the mechanism by which energy is transferred between atoms. Generated EME, rises and falls with the energy in the environmental EME.
[second law of thermodynamics]

The natural (minimum entropy) state of the universe is the collection of all Quanta into a single entity through magnetism; the ultimate-body,
or,
after all universal energy has been lost (to space outside our universe), all universal particles have reverted to lone electrons and protons, and electrons no longer possess the kinetic energy to orbit a proton.
[third law of thermodynamics].

We perceive these energies as described below:

> Mass is the resistance we feel in trying to move or deviate viscous matter.

> Gravity is the physical attraction between atomic particles.

> Light is a range of electro-magnetic energy that we can see with our eyes.

> Heat is the electro-magnetic energy we feel through our atoms.

> Viscous matter is that we can handle.

> Gaseous matter is that we breathe and keeps us warm.

This relates to our experiences as described below:

Low-temperature scenario: when you see an object, such as a cup, you are seeing all the adjacent atoms in that cup held together with magnetic field energy. In this form, the atoms are sufficiently close together to prevent the atoms in, say, your hand, from passing between the atoms in the cup, allowing you to touch but not penetrate the cup.

The weight you feel when you lift the cup, is created by the magnetic energy between the Quanta in the cup and those in the earth.

High-temperature scenario: If sufficient electro-magnetic energy (heat) is trapped by the electrons in the cup and your hand, the electrical charge energy in all the protons will exceed the magnetic field energy forcing the atoms in the cup and your hand to repel each other and intermingle, in a form that we understand as gas.

Energy cannot be lost or gained, and it can only be transferred by electro-magnetic radiation.

2.1 Distance & Time

In physics, distance (d) is the length of path between two points; either straight or circuitous. And time (t) is the passage of time between events.

Distance and time are two of the only four variables required to explain the only properties of every branch of science and engineering; energy [kg.$(m/s)^2$ & C.$(m/s)^2$].

Neither universal time or distance deform around massive celestial bodies, they are both universal constants;

> A unit of distance is always exactly the same, anywhere and everywhere throughout the universe

> A unit of time is always exactly the same, anywhere and everywhere throughout the universe

Distance and time are used to define velocity (v = d/t) and acceleration (a = d/t^2)

2.2 Charge (magnetic & electrical)

Magnetic and electrical charges are the other two variables required to explain the only properties of every branch of science and engineering; energy [$\mathbf{kg}$.$(m/s)^2$ & $\mathbf{C}$.$(m/s)^2$].

Every lone electron possesses exactly the same electrical (e) and magnetic (m) charge of exactly the same magnitude, everywhere in the universe.

Every lone proton possesses exactly the same electrical (e) and magnetic (m) charge of exactly the same magnitude, everywhere in the universe.

Along with distance and time, electrical and magnetic charges, generate all universal energy;

> electrical energy; E_E = e.v.d

> magnetic energy E_M = m.v.d

2.3 Electro-Magnetic (EME)

EME is the electro-magnetic field energy generated and radiated by rotating an electro-magnetic charge about another. Because it is generated by a proton-electron pair, its shape will be an helix, varying between plus and minus electrical charge and minus and plus magnetic charge.

Whilst its wavelength, amplitude and frequency vary with temperature, it always travels at exactly the same velocity; (299792459 m/s), which we refer to as the *speed of light*, but is of course, the same speed for *all* EME.

EME is emitted in a direction normal to the plane of a proton-electron pair according to the '*right-hand-rule*', and will travel in a straight line until deflected by magnetism, reflection or diffraction.

EME wavelength, amplitude and frequency are responsible for the electro-magnetic spectrum (light, radio, X, γ, etc.), which is defined by an orbiting electron's kinetic energy. Its intensity (brightness) is defined by the concentration of emanations. Calcium, for example, is an atom with twenty proton-electron pairs and will therefore emit EME in twenty directions. Millions of such atoms will emit EME in millions of directions.

The EME generated by a proton-electron pair rises and falls with the kinetic energy (velocity) in the electron, which rises and falls with the energy in the radiation feeding it. In this way, [heat] energy in matter naturally stabilises with the energy in its environment.

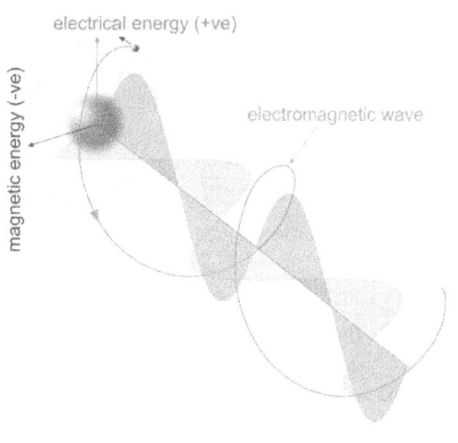

electrical energy (+ve)

magnetic energy (-ve)

electromagnetic wave

If a proton-electron pair falls below temperature T_x, the proton will no longer be able to hold onto its electron partner, which will fly off on its own at 17162.242521927m/s. This pair will no longer radiate EME; they will exist as lone particles.

However, the EME in outer space is at ≈2.7255 K, so, T_x is not possible in our universe, even in outer space.

On reaching temperature T_n, the orbiting electron's velocity will achieve light-speed (c). Immediately this is achieved, the electron will unite with its proton partner to create a neutron. This pair will no longer radiate EME, but it will store the energy it was generating at the moment of union.

2.3.1 EME Energy

There are two principal means of measuring EME energy; Boltzmann and SHC, but they appear to generate different temperatures, for example, a proton-electron pair @ 300K;

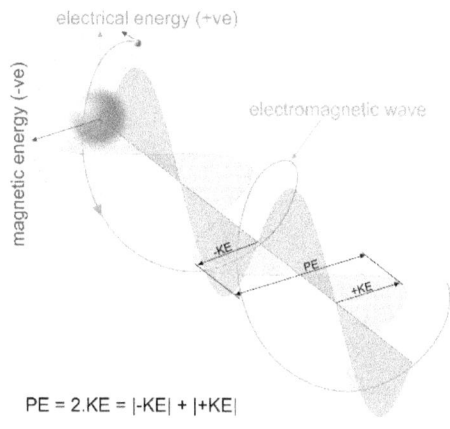

Boltzmann:

$E_B = k_B.Y.\underline{T} = 3.94042969432577E\text{-}20$ J

SHC:

$E_S = SHC.Y.m\square.\underline{T} = 1.9702148471629E\text{-}20$ J

$E_B = 2.E_S$!

However, EME has two measurable energies; *magnitude* (KE) and *range* (PE).

Magnitude is a measure of the *kinetic* energy in the orbiting electron generating the EME, and is used in the calculation of specific heat capacity; SHC = J / K.kg, where 'J' refers to EME magnitude.

Range is a measure of the total energy in EME; $E = |E^-\square\square| + |E^+\square\square|$, and is that defined by Boltzmann in his calculation; k_B = J/K, where 'J' refers to *potential* energy[#] in the proton-electron pair generating the EME.

When quoting EME energy, it is usual to specify its *magnitude*, rather than its *range*, but they both equally apply.

Therefore, both calculations are correct because they give the same temperature:

$$\underline{T} = KE / SHC.Y.m\square = PE / Y.k_B = 300 \text{ K}.$$

[#] PE = 2.KE in circular orbits such as this in proton-electron pairs (chapter 2.5.7)

Electro-magnetic radiation always travels at the same velocity, which we currently refer to as the speed of light: c = 299792459 m/s, but is of course, the same speed as all electro-magnetic energy.

Knowing this, and that its energy is exactly the same as the kinetic energy in the electron that transferred it, we can calculate its other properties (chapter 2.5.3.3).

2.3.2 EME Deflection

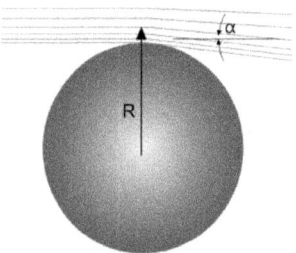

EME is emitted by a proton-electron pair in a direction normal to the plane of the orbiting electron according to the 'right-hand-rule', and will travel in a straight line until deflected by magnetism, reflection or diffraction.

EME is deflected by bodies of magnetic charge according to Newton's gravitational constant (G). This deflection becomes significant as it passes massive bodies such as force-centres and stars.

$\alpha = \text{Atan}(4 . a_o/R . m_s/m_u)$ {m̶/m̶ . k̶g̶/k̶g̶}

Where:
$G = a_o.c^2/m_u$ {m³ / kg.s²}
m_s = the mass of our sun {kg}
m_u = unit mass of ultimate density {kg}
R = distance from the centre of mass {m}
a_o = Rydberg's radius {m}

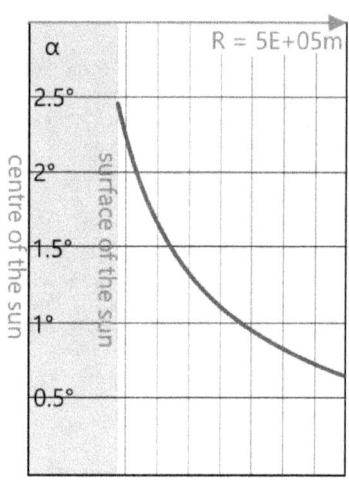

Because EME radiated by the proton-electron pairs in the innermost atoms within a body gets bounced around by its neighbouring pairs, it will take much longer for core temperatures to stabilise. This is why the surface of a body cools faster than its centre. It is also why EME appears to slow down when passing through matter such as water or glass. In practice, the EME does not slow down, it simply gets bounced around.

2.4   Heat & Light

Heat and light are the means whereby we (animals) sense EME:

Heat is how we sense electro-magnetic energy via our body's molecules.

Light is how we sense a small band of the electro-magnetic spectrum through our eyes.

It is because of the heat and light radiated by our sun that we know EME cannot possess mass. If light were photons, radiating energy at its current magnitude, our sun would have lost all of its electrons in the first few seconds of its existence; it would have reverted to a cloud of lone protons and would no longer be able to hold on to its satellites (planets).

We can now conclude that **all heat is radiated**, even within viscous matter. Conduction is simply a term used to describe heat transfer within viscous matter, but the means whereby heat is transferred between its atoms is via EME.

Also, **convection** is not the transfer of heat upwards; heat (EME) is radiated equally in all directions. What we understand as convection is simply the atoms holding the highest temperature - i.e. those with the greatest repulsive e' - are simply looking for space, pushing neighbouring atoms away. This will inevitably cause them to move upwards; away from a planet's magnetic attraction (gravity).

2.4.1 Heat

Heat is the EME generated by the kinetic energy in an atom's electrons. The heat we *feel* is from the senses we have developed to tell us when this kinetic energy is too high or too low. Electron kinetic energy is generated by the EME it absorbs from its surroundings.

It is important to remember that all the EME generated in the universe is just that; EME. It possesses; no light or heat - nothing, apart from energy, and the entire spectral band is simply a single range of EME from 3.17665E+09Hz to 1.6932E+22Hz. We [humans] have split the spectrum into special bands; "γ, X, ultra-violet, light, infra-red, micro, radio" for our own convenience, these bands will mean nothing to any other form of life.

If you or I, devoid of electrons - impossible I know, but bear with me - were to sit in the space between the sun and the earth, we would not be able to detect the sun's radiated EME. It would be invisible in every sense to the fictitious you (or me). EME is useless to all forms of life unless it can be detected.

Whilst EME doesn't deteriorate with distance travelled, we don't feel the sun's surface temperature (5778K) here on Earth because the energy *density* (Joules per square metre) radiated at the sun's surface is distributed over a spherical surface area between 45,000 and 48,000 times greater (dependent on the time of year). Therefore, the EME *density* we receive will be correspondingly less.

Life here on Earth, has evolved to detect and use this energy through our complex molecules. The trouble is, such molecules have energy tolerance levels, outside which they would no longer function; i.e. their state-of-matter, strength or condition (gas-viscous) could change, or inter-atomic bonding could fail.

If a block of viscous iron, the function of which is to be viscous, received sufficient EME to increase its proton electrical charge energy above that of its atomic magnetic field energy, it would become a gas. And it would cease to be *a block of iron*; i.e. no longer functional.

You can't damage a block of iron, so it doesn't need senses. It doesn't matter how many times you change it from gas to viscous and back again its elements always remain iron. The higher its temperature the stronger its atoms remain, until the innermost electrons achieve the *speed of light*, when it will become a different element (Z-1 or Z-2).

We (humans) have five senses - if you exclude time - smell, touch, taste, sight and hearing; each of which were developed for our use and protection.

There are tolerance levels regarding acceptable amounts of EME any living organism can receive and remain functional. Therefore, all living organisms have developed senses, that can be used to ensure that these tolerance levels are maintained.

All the EME in our environment is shared between all of our electrons. The greater the EME *density*, the greater the *heat* we feel. Irrespective of the *temperature* of the atoms that generated the EME, if the energy *density* can be shared throughout all the electrons in our body without exceeding its tolerance levels, we will remain functional. For example:

200,000 calcium atoms at 5778K will radiate 4.446E-13 J of EME.

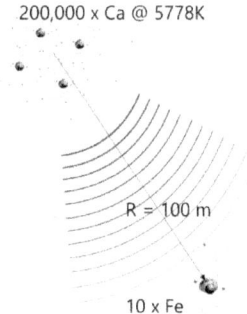

200,000 x Ca @ 5778K

R = 100 m

10 x Fe

A body of 10 iron atoms with a surface area 100th that of the calcium atoms, 100 metres away would absorb a total of 3.54E-20 J of heat energy, resulting in a temperature rise of 9K in the iron.

In other words; it is not the *temperature* of the atoms emitting the EME that defines our body-temperature; it is the quantity of EME absorbed by our body's electrons.

2.4.2 Light

We humans have designated as light EME with a wavelength range between 4E-07m (blue) to 8E-07m (red). We (humans) only call it light because we can *see* it with an unaided eye.

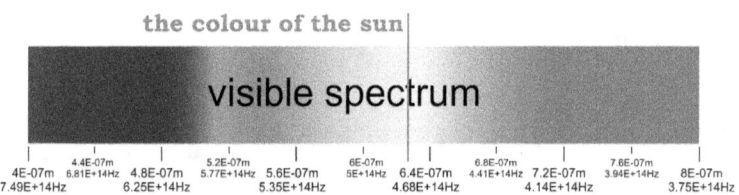

Ultra-violet and infra-red are also EME ranges, but we don't call them light because we can't *see* such wavelengths without artificial aids.

Question: *"If colour is defined by EME, and the surface of the sun is at a temperature of 5788K and looks yellow; how can my towel (at 300K) also look yellow?"*

Colour is a range of EME wavelengths that we cannot detect until it is absorbed by the electrons in our optical receptors (eyes).

Light is the intensity of visible EME wavelengths, or put simply; the number of EME rays per square metre.

We (humans) have developed eyes to detect a particular bandwidth of EME that best suits our purpose. Other lifeforms have developed receptors that best suit their own environment that may be outside the aforementioned *optical* range. Irrespective of a lifeform's preferred optical bandwidth, the purpose of sight is the same; to see what's in its environment and how best to exploit it.

Every proton-electron pair in the universe, *including that block of wood in the garden*, emits EME at a wavelength commensurate with the kinetic energy in its electrons. But you cannot see the EME radiated by that block of wood here on Earth, because it is radiated in the infra-red range. This is the reason infra-red cameras reveal objects in the dark here on Earth. They are actually capturing the electro-magnetic energy given off by the objects themselves, not the electro-magnetic energy radiated by the sun.

Unlike in a prism, the diffraction of light through natural matter is not organised. The image of that block of wood, is the sun's EME reflected and/or refracted by or through its constituent atoms and molecules.

Just as with heat, if the EME received by your eyes is not too intense, i.e. its density remains within your body's tolerance levels, the sun's rays you see will do you no harm.

THE PROTON-ELECTRON PAIR

Whilst our sun's *atmosphere* comprises proton-electron pairs as hydrogen and helium due to its heat, its internals comprise all of nature's elements, the proton-electron pairs of which give off EME according to their shell numbers. A bright star therefore emits a preponderance of electro-magnetic radiation due to the temperature of its surface proton-electron pairs (helium and hydrogen @ 5788K). But its internal matter is radiating EME from all of its electron shells @ temperatures from $\underline{T}\square$ to 125K, which is why distant stars look white.

So, when you see a *yellow* towel here on earth, your eyes are detecting the EME radiated by the sun at a wavelength of ≈6.3E-07m (yellow) diffracted by the molecules in your towel, but at an intensity that will not harm your eyes.

2.5 How it Works

2.5.1 Force

Force is the effort required to induce acceleration in quanta, either by changing velocity or direction.

$$F = m.a$$

Gilbert, Newton & Coulomb generated the mathematical formulas for the potential force between bodies:

Gilbert/Newton: $F_M = G.m_1.m_2/d^2$
Coulomb: $F_E = k.e_1.e_2/d^2$

It is important to understand that because magnetic energy accrues, m_1 and m_2 may be any two different values. In the case of electrical force, however, because electrical charge is shared, 'e_1 & e_2' must be equal to the lesser of the two values. If for example e_1 is greater than e_2, only e_2 will be shared between the two, and electrical charge 'e_1-e_2' will be left over for other work. In this case, his formula will look like this:

Coulomb: $F = k.(e_2/d)^2$

The ratio between the two forces (magnetic and electric) in a proton-electron pair is the coupling ratio:

$$F_E = k.(e/d)^2$$

$$F_M = G.m_e.m_p/d^2$$

$$\varphi = F_M/F_E = 4.40742111792334E\text{-}40$$

2.5.2 Torque & Moments

Torque and moment are both simply mechanical forms of energy: $T = M = F.d$, where 'd' is the distance at which the force is applied.

It is a means whereby a fixed input force can be made to alter an output force, simply by changing the relative distances at which the forces are applied. They are both effectively levers.

2.5.3 Energy

Energy is the distance over which a force is applied. It appears in three principal forms:

Static: potential energy

Atomic: $E_E = k.e^2/d$

$E_M = G.m_e.m_p/d$

Inter-atomic: $PE = m.a.d$

Dynamic: kinetic energy

$KE = \frac{1}{2}.m.v^2$

Transfer: electro-magnetic energy (EME):

$EME = h'/d$

Energy is force multiplied by distance; $E = F.d$, where:

Static: 'd' is the distance between the two bodies

Dynamic: 'd' is the distance travelled by the body

Transfer: 'd' is EME amplitude (A)

2.5.3.1 Kinetic

Kinetic [dynamic] energy, which exists in all moving particles, is always positive and induced via electrical, magnetic, electro-magnetic or impact (potential energy).

EME is absorbed by orbiting electrons and converted into kinetic energy.

Kinetic energy in a satellite following a circular orbit (such as in an atom) is not induced into the satellite by its force-centre such as in elliptical orbits; it must be provided by the satellite itself.

Its linear (or curvilinear) mathematical relationship is: $KE = \frac{1}{2}.m.v^2$

The kinetic energy of a satellite orbiting in a circular path is exactly half the satellite's potential energy: $KE = \frac{1}{2}.PE$ (chapter 2.5.7)

Kinetic energy can also be angular (spin).

Spin energy in the proton:

$$SE_p = KE_e + PE = \frac{1}{2}.m_e.v^2 + (\text{-})m_e.v^2$$

$$SE_p = -\frac{1}{2}.m_e.v^2$$

Kinetic energy in the electron:

$$KE_e = \frac{1}{2}.m_e.v^2$$

The total kinetic energy in a proton electron pair is therefore:

$$E = SE_p + KE_e$$

$$E = 0$$

Nature always balances its books!

2.5.3.2 Potential

Potential [static] energy (PE) is the positive or negative, magnetic or electrical attraction between Quanta. It acts between all Quanta permeating the universe, irrespective of separation distance (d).

What we currently refer to as gravity is actually the magnetic potential energy between all universal Quanta.

Negative potential energy, e.g. gravity, holds particles together, and;

positive potential energy, e.g. centrifugal (CE), drives them apart.

Whilst the potential energy radiated by both magnetic and electrical charge retain their magnitude irrespective of distance, it is distributed over the spherical area at that distance. This is why such a potential force between any two bodies (gravity) *appears* to diminish with the square of the distance between them.

Its linear mathematical relationship is:

PE = m.a.d

in which 'a' may be positive or negative.

In a balanced system (e.g. orbits), both PE and CE must be equal. And in circular orbits, such as atomic, potential energy between Quanta is always twice the kinetic energy in the orbiting satellite, so their mathematical relationship is:

$PE = 2.KE = 2.(\frac{1}{2}.m.v^2) = m.v^2$

2.5.3.3 Electro-Magnetic (EME)

Irrespective of wave shape, electro-magnetic EME follow a sinewave profile and together possess the energy of the orbiting electron. EME properties may be calculated like this:

If a shell-1 (N=1) electron temperature (T_1) is known (the measured temperature of matter), the electron temperature (T_N) in any shell may be calculated thus:

$$T_N = T_1/N \qquad K$$

For example; the electron temperature at the Rydberg [orbital] radius (a_o):

$$T = X\square /a_o = 33192.4000063507 \quad K$$

Wavelength (λ) is defined by the orbital velocity (v) of the electron transferring it:

$$\lambda_N = c/f \qquad \{m\}$$
$$\lambda_N = 2\pi R\square . (T\square/T_N)^{1.5} \; \{m\}$$
$$\lambda_N = (R_N/R\square)^{1.5} / R_\infty \qquad \{m\}$$

where: R_N is the electron orbital radius
where $R\square$ is an unknown orbital radius; 8.40016460895157E-11 m

Frequency (f) is defined by the orbital period of the electron transferring it:

$$f = 1/t_N = c/\lambda_N = v / 2\pi.R_N = g / 2\pi.v_N \quad \{H_z\}$$

where: t_N is electron orbital period ($t_N = 2\pi.R/v$)

Amplitude (A) is equal to the orbital radius of the electron radiating it:

$$A_N = R_N \qquad \{m\}$$

Energy (E) may be determined using the modified version of Planck's constant (h') or the orbital velocity (v) of the electron transferring it:

$$E_N = h'/A_N \qquad \{J\}$$
$$E_N = \tfrac{1}{2}.m_e.v_N^2 \qquad \{J\}$$
$$E_N = a_o.R_\gamma/A_N \qquad \{J\}$$

Which is also the same energy as the orbiting electron's kinetic energy; KE

The electrical **charge** (e') in EME is:

$$e' = m\square.RC . T/T\square \qquad \{C\}$$

which maximises at: $e_n = m\square.RC$ {C}

2.5.4 Power

Power is the rate of energy expenditure, which is why it is measured in units of Joules per second or foot-pound force per hour.

If we apply this argument to kinetic energy, we get:

$$P = \tfrac{1}{2}.m.v^2/t = \tfrac{1}{2}.m.R^2/t^3 = \tfrac{1}{2}.m \cdot R/t \cdot R/t^2 = \tfrac{1}{2}.m.v.a$$

If we apply this argument to angular (rotational) energy, we get:

$$P = 2\pi N.T = 2\pi/t \cdot F.R = 2\pi \cdot m.R/t^2 \cdot R/t = 2\pi \cdot m.R/t \cdot R/t^2 = 2\pi.m.v.a$$

In both cases, power is momentum ($M = m.v$) multiplied by acceleration ($M = m.v.a$).

2.5.5 Pressure

All pressure, in both states of matter, acts in three dimensions; it is actually force-density, as opposed to mass-density. But we almost always need to consider pressure in one direction when applying it to practical applications, making it force per unit area, instead of force per unit volume.

If we multiply force per unit volume by a unit length, we get:

$$p \; (N/m^3) .1 \; (m) = p \; (N/m^2)$$

It is now in a form we can use.

The pressure in all matter, irrespective of its state, may be calculated either by the universally known and accepted formula ($p.V = n.R_i.T$) …

$$p = R_i.T.\rho / (RAM/1000) = R_a.T.\rho$$

… or like this:

$$p = -PE_1 / Y.d^3 = k_B.T / d^3$$
where 'd' is the mean separation [distance] between adjacent atoms.

All of which give exactly the same results.

2.5.6 Temperature

Temperature is a term of convenience we use to measure the energy of the EME emitted by a proton-electron pair. It was configured by the following scientists in the 19th century:

Kelvin: $T_K = PE / k_B.Y$ {K}

Celsius: $T_C = T_K - 273.15$ {°C}

Rankine: $T_R = 9/5 \times T_K$ {R}

Fahrenheit: $T_F = T_R - 459.67$ {°F}

Because temperature is calculated thus:

$R = X□ /T$

@ zero Kelvin the electron's orbital radius will be infinite (R=∞)

An orbiting electron will have left its proton partner long before this occurs. It can therefore be safely assumed that there is no such thing as zero temperature.

The following Table lists the key universal temperatures:

	Formula:	Temperature (K)	electron velocity (m/s)	orbital radius (m)
		$T = PE/k_B'$	$v = \sqrt{[T/X]}$	$R = X□/T$
abs. minimum	$T□ = X.(c / Y.\xi□.)^2$	2.04274907568265	17162.242521927	8.59854098572228E-07
e' = e	$T□ = T□.e / m□.RC$	339468.852842837	6996268.84651131	5.1741600160694E-12
Rydberg radius	$a□ = R□.(\xi□/4\pi)^2$	33192.4000063507	2187690.35053551	5.2917721067E-11
Planck min.	$v□ = v□/\xi□$	210.193328535837	174090.866621084	8.35643156381571E-09
Planck mean	$v□ = v□ / \sqrt{.\xi□}$	361962.554671561	7224342.80705004	4.85261843362263E-12
Neutronic	$v□ = c$	623316124.717178	299792459	2.81793795383896E-15

Absolute Minimum: the temperature at which an electron will leave its proton partner.

Rydberg Radius: the temperature that gave us the dynamic ratio; $\xi□$ ($T□ = X□/a□$)

Planck Minimum: the gas-transition temperature of the largest noble atom ($T□ = X.v□^2$).

Planck Mean: the mean value between his minimum and the neutronic ($T□ = X.v□^2$).

Neutronic: at which a proton-electron pair will unite as a neutron ($T□ = m.c^2/kB'$).

Given the following:

EME velocity = 299792459 m/s (c)
'Big-Bang' initial velocity ≈ 1772254 m/s (0.5916% x c)

velocity today of all universal matter ≈ 230000 m/s (0.07672% x c)

The EME radiated at the last 'Big-Bang' is today (1.303E+26m) well outside the limits of the universe today (4.353E+23m). The temperature measured during the Cobe project (≈2.7255 K) had nothing to do with heat left over from the last 'Big-Bang', it was simply detecting the EME (heat) radiated by all the stars and planets in the universe.

2.5.7 E=mc²

Henri Poincaré derived his formula at the end of the 19th century, but he wrote it like this; $c = \sqrt{[E/m]}$. However, he was unsure of its meaning.

The difference between kinetic and dynamic energies looks like this (chapter 2.5.3):

$KE = \frac{1}{2}.m.v^2$

$PE = m.a.d$

According to Newton's laws of orbital motion, the potential energy between a satellite and its force-centre in non-circular elliptical orbits looks like this:

$a = v\square.v / d.(1-e)$

But in circular orbits:

$v\square = v$ & $e = 0$, therefore; $a = v^2/d$

Therefore, the potential energy between a force-centre and its satellite in circular orbits is:

$PE = m.(v^2/d).d$

i.e.:

$PE = m.v^2$ & $KE = \frac{1}{2}.m.v^2$

$PE = 2.KE$

Therefore, Poincaré's formula applies to orbiting electrons, and it works like this …

$E = PE = 2.KE = 2 . \frac{1}{2}.m_e.v^2 = m_e.v^2$

… and when the orbiting electron achieves 'light-speed' (c), as is the case for proton-electron pairs in the core of bright stars and planets, this formula becomes:

$E = m_e.c^2$

at which instant, the proton-electron pair unites to create a neutron.

This is the meaning of Henri Poincaré's formula.

Energy and mass do not transpose with velocity.

3   Atomic Particles

There are only two atomic particles in the universe and they are definitive; they do not comprise smaller (sub-atomic) particles. The third particle (the neutron) is not a true particle, it comprises a proton and an electron united through high temperature.

Every electron is exactly the same as every other electron and every proton is exactly the same as every other proton.

3.1 The Proton

A proton is a packet of electro-magnetic charge.

All protons are identical.

A proton's magnetic charge is constant and non-polar.

A proton's electrical charge is positive.

A proton's electrical charge is constant and equal to that of an electron whilst it is not partnered with an electron.

When partnered with an electron, a proton's electrical charge varies with the kinetic energy in its orbiting electron.

The additional electrical charge collected from its electron partner is held and used to repel adjacent protons (atomic nuclei).

Protons have no kinetic energy of their own. They move only when caused to do so by external means.

Protons do not exist in outer space. Their magnetic charge ensures that they are always attracted to the nearest celestial body.

The proton's properties are listed below.

Mass: $m = 1.67262163783E-27$ kg

Electrical charge (alone): $e = 1.60217648753E-19$ C

Electrical charge (with electron partner): $e_\square = m_e.RC . \sqrt{[T/T_n]}$

Electrical charge (maximum): $e_n = m_e.RC = 2.94183820093364E-16$ C

Density: $\rho = 7.1266079635045E+16$ kg/m^3

Radius: $r = 1.77613270336827E-15$ m

Volume: $V = 2.34700946985653E-44$ m^3

Polar moment of inertia: $J = 2.11061258698748E-57$ kg.m^2

Lone protons cannot be coalesced into viscous matter because they all possess a positive electrical charge. They can only exist as a gas.

3.2   The Electron

An electron is a packet of electro-magnetic charge.

All electrons are identical.

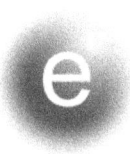

An electrons magnetic charge is constant and non-polar.

An electrons electrical charge is constant and negative.

Electrons only exist in free-flight or as part of a proton-electron pair.

In free-flight, electrons travel at the velocity they had achieved at the moment they escaped their proton-electron partnership. Their kinetic energy will not change until they again become partnered with a proton.

When partnered with a proton, electrons convert the energy they collect from the EME in their surroundings, into kinetic energy.

Electrons do not exist in outer space. Their magnetic charge ensures that they are always attracted to the nearest celestial body.

The electron's properties are listed below.

Mass: m = 9.1093897E-31 kg

Electrical charge: e = -1.60217648753E-19 C

Density: ρ = 7.1266079635045E+16 kg/m³

Radius: r = 1.45046059426276E-16 m

Volume: V = 1.27822236702922E-47 m³

Polar moment of inertia: J = 7.66586456056651E-63 kg.m²

Lone electrons cannot be coalesced into viscous matter because they all possess a negative electrical charge. They can only exist as a gas.

3.3 The Neutron

A neutron is a proton-electron pair united through high temperature (T_n).

All neutrons are identical.

The electrical charge in a neutron is zero.

The magnetic charge in a neutron is constant and non-polar.

Neutrons have no kinetic energy of their own.

Neutrons are only created, and can only exist, within an atom.

Neutrons store all the kinetic, potential and spin energy their component parts possessed at the time of their creation.

The neutron's properties are listed below.

Mass: m = 1.6735325768E-27 kg

Electrical charge: e = 0 C

Density: ρ = 7.1266079635045E+16 kg/m^3

Radius: r = 1.77645508248591E-15 m

Volume: V = 2.34828769222356E-44 m^3

Polar moment of inertia: J = 2.11252872891479E-57 kg.m^2

Stored energy: E = 1.63785606465701E-13 J

Neutrons are not particles in their own right. They cannot be picked up (trapped) by accident or design. They are [only] created inside an existing atom, and they cannot exist outside one. When ejected from their nucleus, they will revert to their original component parts, a proton (alpha-particle) and an electron (beta-particle), or they will revert to a proton-electron pair if trapped within their parent atom.

4   Proton-Electron Pair

A proton-electron pair is a single
proton with a single orbiting electron
partner. The electron's orbital path is
circular because it provides its own
kinetic energy.

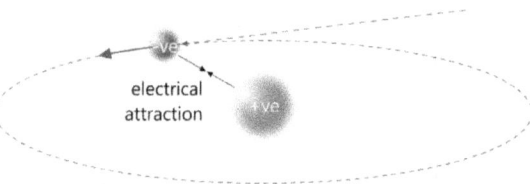

The electron is attracted to its proton
partner due to their opposite electrical charges. But the electron will not attach itself to its
proton because of its intrinsic kinetic energy; it must keep moving.

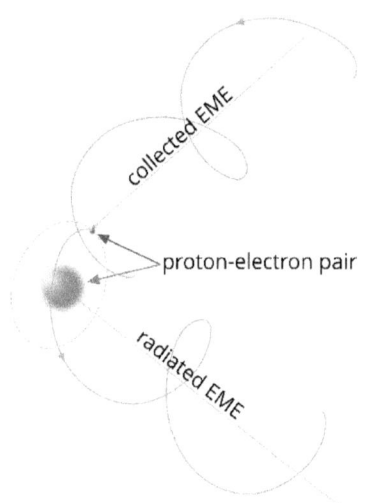

This partnership acts like a generator; a negative
electro-magnetic charge orbiting a positive electro-
magnetic charge will generate electro-magnetic energy
(EME). In other words, the EME generated by a
proton-electron pair is the same magnitude as the EME
feeding it. Only proton-electron pairs emit EME; lone
particles cannot.

The electron converts the EME it collects from its
environment to kinetic energy, which varies with the
EME feeding it. We interpret the EME it collects and
emits as its temperature. The greater the EME, the
higher its temperature, and vice-versa.

In the case of a proton-electron pair, the potential force
holding an electron in orbit about its proton partner is
as coulomb defined it; $F = k.(e/R)^2$, in which 'e' is the elementary charge unit, and the
electron's orbital radius is as Newton defined it;

PE = F.R = 2.KE (chapter 2.5.7)

PE = $k.e^2 / R$

R = $k.e^2 / 2.KE$

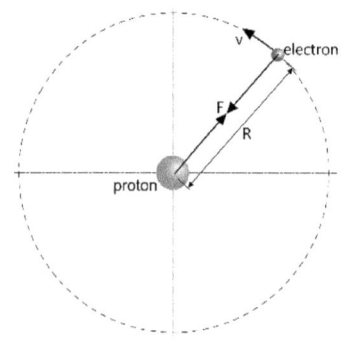

So, Newton's laws of orbital motion show us that the
electron's orbital radius varies inversely with its kinetic
energy. In other words, because the electron's mass
remains constant, as a proton-electron pair's temperature
rises, its velocity increases and its orbital radius reduces,
and vice-versa. I.e. the structural strength of a proton-
electron pair - along with its parent atom - rises and falls contrary to its temperature.

THE PROTON-ELECTRON PAIR

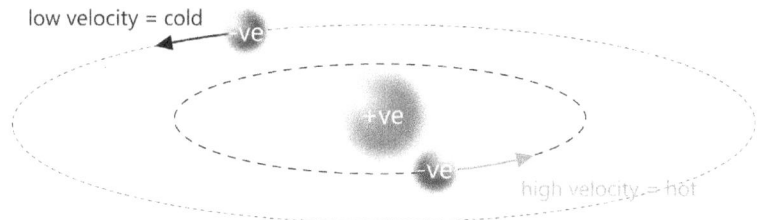

The electrical charge in an electron never changes. That of a lone proton also remains constant and equal to that of the electron, but varies with the EME its electron partner collects as soon as it acquires one. An orbiting electron transfers energy it collects from its environment and passes it on to its proton partner via their opposite electrical charges. The proton uses its excess magnetic charge (m_p-m_e) to hold onto this electrical charge (e□), which it then uses to repel neighbouring protons in the same and adjacent atoms.

Electrical sharing (as opposed to magnetic accruing) is the reason a proton will only host one orbiting electron, despite its additional electrical charge (e□):
Unlike magnetic attraction, according to Newton's force law;

$E = G.m_1.m_2/R$

in which both m_1 & m_2 can be any value, electrical attraction (Coulomb) between two particles of different electrical charge (e_1 & e_2) will be defined by the square of the lowest value:

If $e_1 < e_2$ then; $E = k.e_1^2/R$
If $e_1 > e_2$ then; $E = k.e_2^2/R$

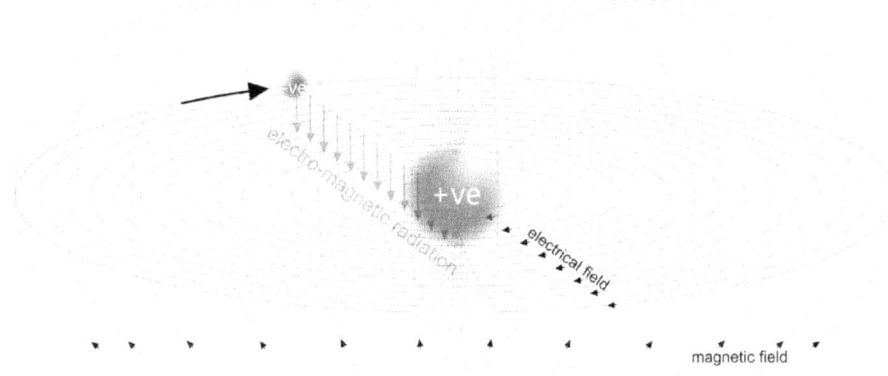

In addition to EME, proton-electron pairs also generate a polar magnetic field - that works contrary to an electrical field - in that it travels from the positive (North) face of the orbital plane and around to the negative (South) face of the orbital plane. It is this force that holds onto neutrons and will ultimately unite the pair as a neutron. It is also the magnetism you see in bar magnets, when an element's shell-1 proton-electron pairs are aligned.

THE PROTON-ELECTRON PAIR

Once trapped, an electron will remain in orbit around its proton partner – even inside an atom - until it is forcibly removed, or its kinetic energy falls below that which its proton can no longer hold onto it. It is then a free electron,

When the orbiting electron achieves the speed of light (c), the magnetic field force will exceed the centrifugal force and the pair will unite as a neutron. This occurs in the core of large celestial bodies where the heat generated due to planetary spin finds it difficult to escape the body's confines, allowing core heat to build up to the neutronic temperature; 623316124.717178 K.

4.1 How it Works

The orbital properties of a proton-electron pair are related to its temperature thus:

Orbital velocity: $v = \sqrt{[T/X]}$

Orbital radius: $R = X\square/T$

where: X and $X\square$ are heat transfer coefficients

All these formulas have been derived from Newton's laws of orbital motion and Newton's and Coulomb's force law.

There are two sets of energy at play in a proton-electron pair;

the magnetic charge pulling the particles together

and

the electrical charge pulling the particles together

both of which are related by the coupling ratio (a constant);

$F_M/F_E = G.m_e.m_p/R^2 \ / \ k.(e\ /R)^2 = G.m_e.m_p \ / \ k.e^2 = \varphi$

therefore, the electrical charge force holding the proton-electron pair together is $1/\varphi$ times greater than the magnetic (gravitational) force. This is how we *know* that our solar system couldn't possibly have accreted from hydrogen gas, because the electrical repulsion force between adjacent hydrogen atoms (both of which are positively charged) is $1/\varphi$ times greater than the magnetic attraction force.

We can calculate the electrical orbital properties according to Coulomb, and the magnetic orbital properties according to Newton, the ratio of which should be $1/\varphi$ for acceleration related values $(/s^2)$ and $\sqrt{[1/\varphi]}$ for velocity related values $(/s)$.

The total magnetic field generated by the proton-electron pair may be calculated thus:

$\mu_p = m_p.R/e^2$

most of which is attributed to neutron partners and adjacent atoms in viscous matter. However, the proportion associated with the orbiting electron;

$\mu = m_e.R/e^2$

generates the attractive field force holding onto the electron may be calculated thus;

$F_M = \mu.g.e^2 \ / \ R$

which breaks down to;

$F_M = m_e.v^2/R$

THE PROTON-ELECTRON PAIR

When F_M exceeds '$F_E = k.(e/R)^2 = G.m\square.m\square / R.\varphi$, the pair will unite as a neutron.

According to Coulomb, a proton-electron pair's electrical performance (electrical charge) can be calculated using the formulas listed in the Tables below.

The total energy (including spin) in a proton-electron pair at 300K is:

$$E_A = -7.88085938865144E\text{-}20 \quad J$$

The kinetic energy in an orbiting electron at 300K is:

$$KE_A = 1.970214847162860E\text{-}20 \quad J$$

Sym	Description	units
m_e	electron mass	kg
m_p	proton mass	kg
T	temperature	K

Table 4.1A: Input Data

Sym	Formula	Description	units
R	$X\square / T$	orbital radius	m
e	0	orbital eccentricity	
A	$\pi.R^2$	orbital swept area	m^2
L	$2\pi.R$	orbital path length	m
K	t^2/R^3	orbital constant of proportionality	s^2/m^3

Table 4.1B: Magnetic Orbital Shape

Sym	Formula	Description	units
v	$\sqrt{[T/X]}$	electron velocity	m/s
t	$2\pi.R/v$	orbital period	s
a	$-v^2/R$	satellite potential acceleration	m/s^2
F	$-k.(e/R)^2$	proton-electron pair potential force	J
PE	$F.R$	proton-electron pair potential energy	J
KE	$-PE/2$	satellite kinetic energy	N
F_c	$m_e.a$	satellite centrifugal force	N
E	$PE + KE$	total energy	J
h	$v.R$	constant of motion	m^2/s

Table 4.1C: Electric Orbital Performance

According to Gilbert & Newton, a proton-electron pair's magnetic performance (magnetic charge) can be calculated using the formulas listed in the following Tables.

Sym	Description	units
m_e	electron mass	kg
m_p	proton mass	kg
T	temperature	K

Table 4.2A: Input Data

Sym	Formula	Description	units
R	$X \square /T$	orbital radius	m
e	0	orbital eccentricity	
A	$\pi.R^2$	orbital swept area	m^2
L	$2\pi.R$	orbital path length	m
K	$\varphi.(2.\pi)^2 / G.m_p$	orbital constant of proportionality	s^2/m^3

Table 4.2B: Magnetic Orbital Shape

Sym	Formula	Description	units
a	$-G.m_p/R^2$	satellite potential acceleration	m/s^2
v	$\sqrt{[a.R]}$	electron velocity	m/s
t	$2\pi.R/v$	orbital period	s
F	$-G.m_e.m_p/R^2$	potential force	J
PE	$F.R$	potential energy	J
KE	$-PE/2$	satellite kinetic energy	N
F_c	$m_e.-a$	satellite centrifugal force	N
E	$PE + KE$	total energy	J
h	$R.v$	constant of motion	m^2/s

Table 4.2C: Magnetic Orbital Performance

The ratio between the two calculation methods; (electrical:magnetic) is thus;

Difference	Symbol	Ratio	
a:a	φ	4.4074211179233E-40	$\ldots/s^2$
v:v	$\sqrt{\varphi}$	2.0993858906650E-20	$\ldots/s$
t:t	$\sqrt{\varphi}$	2.0993858906650E-20	$\ldots/s$
KE:KE	φ	4.4074211179233E-40	$\ldots/s^2$
PE:PE	φ	4.4074211179233E-40	$\ldots/s^2$
F:F	φ	4.4074211179233E-40	$\ldots/s^2$
$F_c:F_c$	φ	4.4074211179233E-40	$\ldots/s^2$
E:E	φ	4.4074211179233E-40	$\ldots/s^2$
h:h	$\sqrt{\varphi}$	2.0993858906650E-20	$\ldots/s$
Table 4.3: Magnetic:Electrical Performance Ratios			

Within an atom, the electrical properties dominate the performance of its particles by the coupling ratio (φ). The electrical properties within a proton-electron pair are therefore calculated thus:

Sym	Formula	Description	units
V	$\underline{T}.m_e / X.e$	voltage	J/C
I	$e.a/v$	current	C/s
Ω	V/I	resistance	$J.s/C^2$
ρ	$\Omega.R$	resistivity	$J.s.m/C^2$
$e\square$		proton electrical charge	C
μ	$m_p.RC . \underline{T}/\underline{T}\square$	magnetic field	$Kg.m/C^2$
λ	$m_p.R/e^2$	EME wavelength	m
A	f/c	EME amplitude	m
f	R	EME frequency	/s
EME	$V / 2\pi R$	EME (energy)	J
	$h'/A = \tfrac{1}{2}.m_e.v^2$		
Table 4.4: Electrical Performance			

4.2   Example Calculations

The heat radiated from matter is the electro-magnetic energy (EME) radiated by all of its atomic proton-electron pairs. Its measured temperature, however, is that generated by the proton-electron pairs in its atom's innermost shells (Shell-1).

The Tables in this Chapter list performance of proton-electron-pairs at nature's key temperatures along with those of our earth and our sun.

Lowest natural: $T_x = X.\sqrt{[c / Y.\xi\square]} = 2.04274907568265$ K

Planck's minimum: $T_o = X.(c/\xi\square)^2 = 210.193328535837$ K

Planck's mean: $T_m = X.c^2/\xi\square = 361962.554671561$ K

Neutronic: $T_n = m_e.c^2 / k_B.Y = 623316124.717178$ K

Earth's atmospheric: $T_a = 300$ K

Sun's atmospheric: $T_S = 5788$ K

4.2.1 Minimum Natural Temperature ($\underline{T}_x$)

The minimum naturally occurring temperature below which the proton can no longer hold onto its electron partner.

Sym	Magnetic Values	Electrical Values	Units
Orbital Dimensions (N/A)			
$\underline{T}$		*2.04274907568265*	K
R		8.59854098572228E-07	m
t		3.14797010446892E-10	s
K		0.15587874533403	s²/m³
Masses			
m_1		*1.67262163783E-27*	kg
m_2		*9.1093897E-31*	kg
Satellite Performance			
v	3.6030169802705E-16	17162.2425219271	m/s
g	1.5097597815343E-25	3.4254947306819E+14	m/s²
F	-1.3752990203382E-55	-3.1204166417078E-16	N
PE	-1.1825564994002E-61	-2.6831030386254E-22	J
KE	5.912782497001E-62	1.3415515193127E-22	J
E	-5.912782497001E-62	-1.3415515193127E-22	J
h	3.0980689177109E-22	1.4757024573170E-02	m²/s
e□	3.535610277366E-40	1.6841164328517E-20	kg - C
Electro-Magnetic Emission			
f		3.176650243852E+12	Hz
λ		9.43737698477225E-02	m
A		8.59854098572228E-07	m
E		1.38042005551962E-20	J
e□		9.64107461280872E-25	C

The electrical charge in the proton partner at this temperature is 'Y' times less than the elementary charge unit (e' = e/Y).

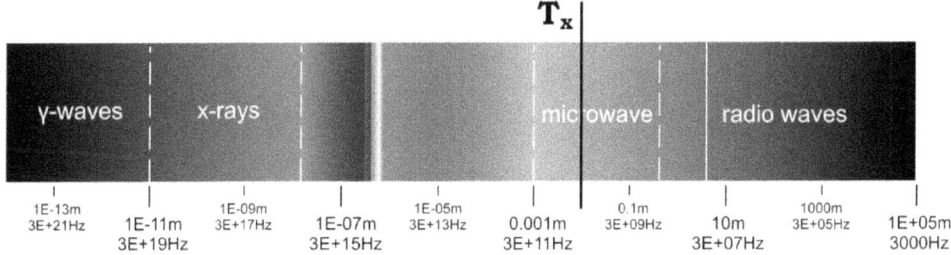

The above image identifies where the electro-magnetic emission radiated by the proton-electron pair at the above temperature occurs in the spectrum

41

4.2.2 Planck Minimum Temperature (T_o)

Planck's minimum temperature.

Sym	Magnetic Values	Electrical Values	Units
Orbital Dimensions (N/A)			
T		*210.19332853584*	K
R		8.3564315638157E-09	m
t		3.0159541991653E-13	s
K		0.15587874533403	s²/m³
Masses			
m₁		*1.672621637830E-27*	kg
m₂		*9.109389700000E-31*	kg
Satellite Performance			
v	3.654839090779E-15	174090.86662108	m/s
g	1.598511120145E-21	3.6268626876711E+18	m/s²
F	-1.456146073318E-51	-3.303850561039E-12	N
PE	-1.216818500860E-59	-2.760840111039E-20	J
KE	6.084092504302E-60	1.380420055520E-20	J
E	-6.084092504302E-60	-1.380420055520E-20	J
h	3.054141273886E-23	1.454778412804E-03	m²/s
e□	9.1093897E-31	1.602176487530E-19	kg - C
Electro-Magnetic Emission			
f		3.315700219442E+12	Hz
λ		9.0416032559912E-05	m
A		8.356431563816E-09	m
E		1.38042005551962E-20	J
e□		9.92040377182125E-23	C

The electrical charge in the proton partner below this temperature is less than the elementary charge unit (e' ≤ e).

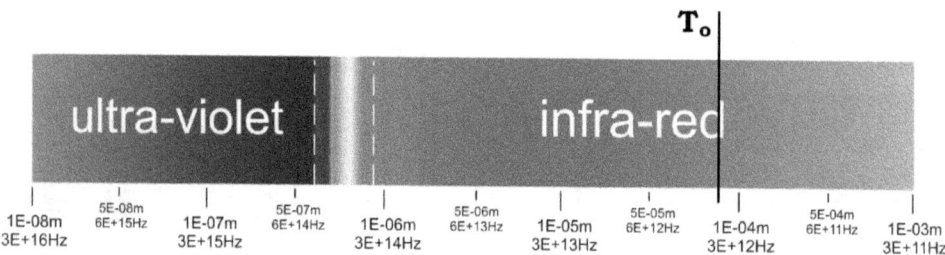

The above image identifies where the electro-magnetic emission radiated by the proton-electron pair at the above temperature occurs in the spectrum

4.2.3 Planck Mean Temperature ($\underline{T}_m$)

Planck's mean temperature.

Sym	Magnetic Values	Electrical Values	Units
Orbital Dimensions (N/A)			
$\underline{T}$		*361962.55467156*	K
R		4.8526184336226E-12	m
t		4.2204393752927E-18	s
K		0.15587874533403	s²/m³
Masses			
m₁		*1.672621637830E-27*	kg
m₂		*9.109389700000E-31*	kg
Satellite Performance			
v	1.516668335845E-13	7224342.80705	m/s
g	4.740292014340E-15	1.0755250944965E+25	m/s²
F	-4.318116725043E-45	-9.797377217898E-06	N
PE	-2.095417281848E-56	-4.754293328873E-17	J
KE	1.047708640924E-56	2.377146664436E-17	J
E	-1.047708640924E-56	-2.377146664436E-17	J
h	7.359812724212E-25	3.505697907630E-05	m²/s
e⬚	9.1093897E-31	6.648638385209E-18	kg - C
Electro-Magnetic Emission			
f		2.369421548510E+17	Hz
λ		1.265255898379E-09	m
A		4.852618433623E-12	m
E		2.37714666443636E-17	J
e⬚		1.70833904084141E-19	C

This temperature occurs between Planck's minimum and the neutronic temperatures.

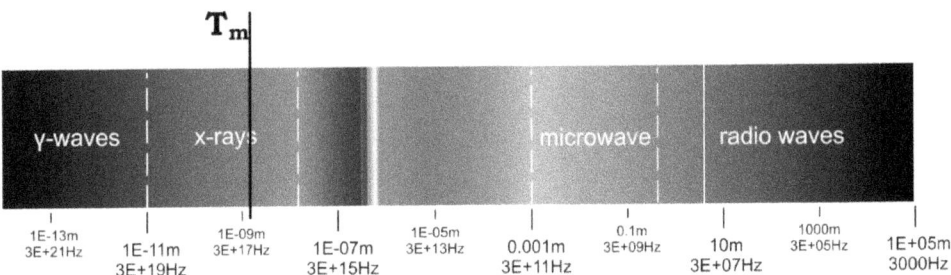

The above image identifies where the electro-magnetic emission radiated by the proton-electron pair at the above temperature occurs in the spectrum

4.2.4 Neutronic Temperature (T_n)

This neutronic temperature; at which the proton-electron pair unite to become a neutron.

Sym	Magnetic Values	Electrical Values	Units
Orbital Dimensions (N/A)			
T	*623316124.71718*		K
R	2.8179379538390E-15		m
t	5.9059612130219E-23		s
K	0.15587874533403		s^2/m^3
Masses			
m_1	*1.672621637830E-27*		kg
m_2	*9.109389700000E-31*		kg
Satellite Performance			
v	6.293800585524E-12	299792459	m/s
g	1.405706103514E-08	3.1894072880784E+31	m/s^2
F	-1.280512470057E-38	-2.905355389913E+01	N
PE	-3.608404689739E-53	-8.187111222625E-14	J
KE	1.804202344869E-53	4.093555611313E-14	J
E	-1.804202344869E-53	-4.093555611313E-14	J
h	1.773553954384E-26	8.447965484908E-07	m^2/s
e□	9.1093897E-31	2.759021413766E-16	kg - C
Electro-Magnetic Emission			
f	1.693204482608E+22		Hz
λ	1.770562634810E-14		m
A	2.817937953839E-15		m
E	4.09355561131267E-14		J
e□	2.94183820093364E-16		C

This temperature is the highest possible in nature.

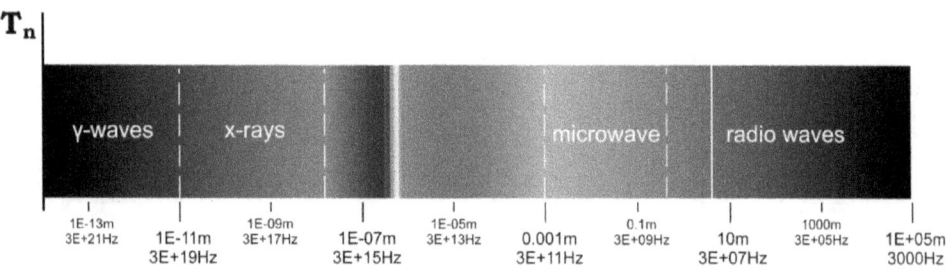

The above image identifies where the electro-magnetic emission radiated by the proton-electron pair at the above temperature occurs in the spectrum

4.2.5 Earth Ambient Temperature (T_a)

The ambient temperature at the surface of the earth in temperate zones ($\approx$27°C); 300 K

Sym	Magnetic Values	Electrical Values	Units
Orbital Dimensions (N/A)			
T		*300*	K
R		5.8548872169345E-09	m
t		1.7687695423507E-13	s
K		0.15587874533403	s²/m³
Masses			
m₁		*1.672621637830E-27*	kg
m₂		*9.109389700000E-31*	kg
Satellite Performance			
v	4.366358844837E-15	207982.67075397	m/s
g	3.256269310661E-21	7.3881510832249E+18	m/s²
F	-2.966262611896E-51	-6.730154737957E-12	N
PE	-1.736713304846E-59	-3.940429694326E-20	J
KE	8.683566524232E-60	1.970214847163E-20	J
E	-8.683566524232E-60	-1.970214847163E-20	J
h	2.556453858519E-23	1.217715080341E-03	m²/s
e□	9.1093897E-31	1.914086312299E-19	kg - C
Electro-Magnetic Emission			
f		5.653647782011E+12	Hz
λ		5.302637705056E-05	m
A		5.854887216935E-09	m
E		1.97021484716286E-20	J
e□		1.41589704691265E-22	C

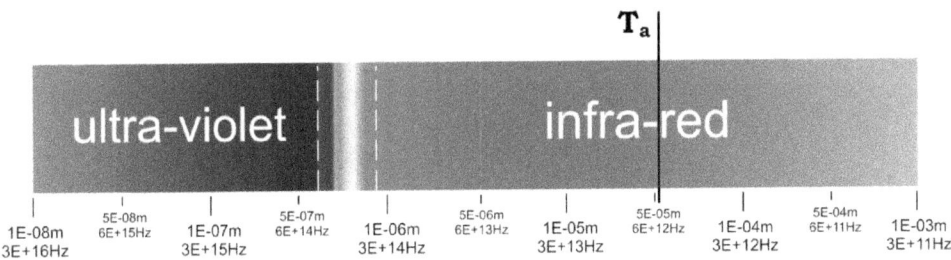

The above image identifies where the electro-magnetic emission radiated by the proton-electron pair at the above temperature occurs in the spectrum

4.2.6 Sun Ambient Temperature (T_S)

This is the temperature at the surface of the sun; 5778K, the reason we can see the colour of the surface of the sun is because its gases are proton-electron pairs, not lone protons (H^+).

Sym	Magnetic Values	Electrical Values	Units
Orbital Dimensions (N/A)			
T	5788		K
R	3.0346685644097E-10		m
t	2.08718213892946E-15		s
K	0.15587874533403		s²/m³
Masses			
m₁	1.672621637830E-27		kg
m₂	9.109389700000E-31		kg
Satellite Performance			
v	1.917887191548E-14	913546.76625580	m/s
g	1.21208995361541E-18	2.75011150780736E+21	m/s²
F	-1.10413997389377E-48	-2.50518374430718E-09	N
PE	-3.35069886948358E-58	-7.60240235691919E-19	J
KE	1.67534943474179E-58	3.80120117845959E-19	J
E	-1.67534943474179E-58	-3.80120117845959E-19	J
h	5.82015197025069E-24	2.77231165367462E-04	m²/s
e☐	2.73173736917685E-21	2.73173736917685E-21	kg - C
Electro-Magnetic Emission			
f	4.79114870402691E+14		Hz
λ	6.25721465810542E-07		m
A	3.0346685644097E-10		m
E	3.80120117845959E-19		J
e☐	2.73173736917685E-21		C

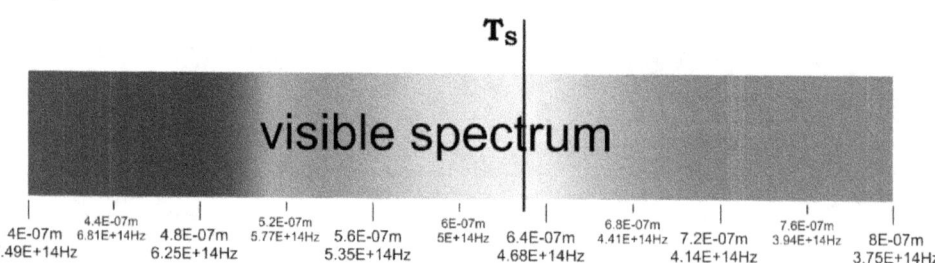

The above image identifies where the electro-magnetic emission radiated by the proton-electron pair at the above temperature occurs in the spectrum

5   The Atom

The atom according to Newton and Coulomb is a system that works perfectly, it needs no sub-atomic particles to hold it together, and every proton, electron and neutron is identical. Uniqueness and uncertainty are unnecessary. Moreover, apart from Planck's contribution, everything needed to resolve the atom completely and accurately was available before the beginning of the twentieth century.

Because all protons repel each other due to their identical electrical polarity, every proton-electron pair within an atom must hold onto at least one neutron isolating it from its neighbouring protons. Proton-electron pairs are the most basic form of matter; hydrogen gas (H), which exist in three forms:

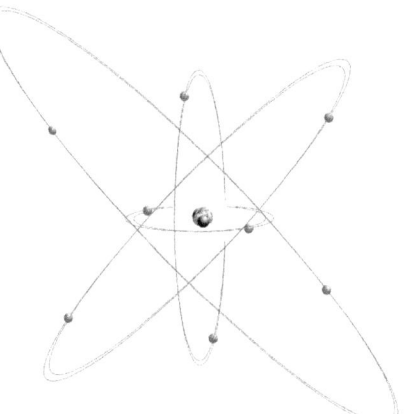

Hydrogen (H); a proton-electron pair.

Deuterium (D); a proton-electron pair with a neutron attached.

Tritium (T); a proton-electron pair with two neutrons attached.

All atoms are created through fusion in the universe's largest massive cold bodies; the ultimate body, the great attractor and galactic force-centres. And all neutrons are created by fissionable energy (Chapter 5.7), converting Shell-1 proton-electron pairs into neutrons that are subsequently trapped by adjacent deuterium atoms.

An atom is therefore a collection of deuterium and tritium atoms in which all the proton force-centres are assembled inside a cloud of orbiting electrons, each of which orbits its proton partner.

Whilst natural hydrogen at the surface of a planet comprises mostly lone protons (H+), the majority of hydrogen atoms at the surface of bright stars and planets are proton-electron pairs (H) as they are the residue of total fissionable dissemination.

You will frequently see references to hydrogen atoms existing in pairs (H_2), this is of course impossible for 99.97% of atmospheric hydrogen (H+) as all lone protons possess positive electrical charge and generate no magnetic field. Hydrogen atoms (H), however, do generate magnetic fields and therefore could (theoretically) become diatomic, and perhaps even exist as a liquid, but the temperature at which this could occur must be very low indeed, e.g. that of outer-space, and can only apply to less than 0.03% of atmospheric hydrogen, and only if at least one of the atoms was deuterium. It is for this same reason that it is impossible that our sun and its solar system accreted from hydrogen.

An atom's atomic number (Z) is defined by the number of protons in its nucleus. This number distinguishes its primary character.

Natural atoms have atomic numbers between 1 (hydrogen) and 92 (uranium). The only apparent exception to this rule is technetium (Z = 43).

Unnatural atoms, i.e. those with atomic numbers greater than 92 (and 43) can be created artificially but they are very unstable. An unstable atom is one that continuously and regularly ejects its neutronic particles (protons and electrons) or breaks apart due to excess neutron-neutron interaction.

Atoms with an atomic number greater than one are created by fusion; i.e. forcing the nucleus of one atom inside the electron shells of another using magnetic (potential) energy. No two atoms can be fused together unless all the protons in both atoms are protected by at least one neutron. Moreover, fusion can only occur naturally in massive cold bodies. Fusion in stars is impossible for two reasons:

1) stars have insufficient mass to generate the core pressure required, and
2) stars are hot, inter-atomic repulsion prevents fusion from occurring

The total kinetic energy in all the electrons in any atom is its quantity of heat. The highest kinetic energy, is in the electrons orbiting in an atom's innermost shell (shell-1).

All of the proton-electron pairs in any atom collect and radiate EME from their surroundings. The range of visible EME emitted by all of the proton-electron pairs within an atom are its Balmer lines.

5.1 Nucleus

The nucleus of an atom contains proton partners together with their attached neutron(s) that are held by the magnetic field energy generated by the proton-electron pairs.

Their positive electrical charges (e') prevent nucleic protons - and their neutron partners - from sitting together in the atomic nucleus. Instead, they will organise themselves, spaced apart, by balancing their electrical charges - both repulsion and attraction - within the innermost orbiting electrons, settling at their lowest energy condition, creating each atom's unique lattice structure.

Nucleic organisation (lattice arrangement) is defined by the electrical isolation of its proton's positive charges, which is determined by its attached neutrons. Neutrons are therefore responsible for an atom's lattice structure. The arrangement of any atomic nucleus is replicated in the lattice structure of the atoms of elemental matter in both viscous and gaseous conditions.

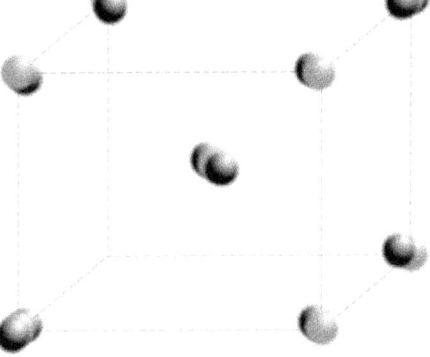

Because of the replication of atomic nucleus and elemental lattice structures, their mathematical relationships are identical.

Regardless of the nucleic pattern, structural integrity generally reduces with increasing nucleic size (i.e. as the atomic number increases), making larger atoms generally (but not necessarily) more unstable due to the greater potential for neutron-neutron interaction, resulting in radioactivity. Technetium appears to be a special case in which the nucleic arrangement of 43 protons is especially vulnerable to neutron decay.

5.2 Electron Shells

All electron shells are circular, equally spaced, and - apart from the hydrogen atom and the outer shell of atoms with an odd number of electrons - *contain two electrons*. This *statement* is slightly misleading, however; because each electron orbits its proton partner, the two electrons in any given shell cannot follow the same circular path. What this *statement* actually means is that they both orbit their proton partners at the same orbital radius, but at opposite sides of the nucleus.

Shell spacing is defined by a balancing act between the repulsion forces in adjacent electrons and the attraction forces in their proton partners, in exactly the same way as Newton described the balance between centrifugal force and potential force in an orbiting satellite and its force-centre.

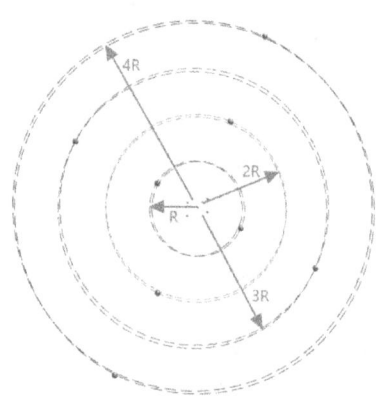

When an inner-shell electron is lost, all other electrons redistribute to fill the gap. Redistribution occurs automatically as electrical forces balance. Shell spacing is equal to the radius of the innermost shell (shell-1).

The orbital radii of all atomic electrons decrease simultaneously as the surrounding EME rises, and increase simultaneously as the surrounding EME falls.

Shell radius is determined by electron velocity, which is determined by its kinetic energy, which is determined by the EME it collects from its surroundings. The greater an electron's velocity, the smaller its orbital radius, and vice-versa. An electron cannot collect more EME than its balancing forces permit. Shell-1 holds the most energy and shell-N (outermost shell) holds the least.

We measure electron kinetic energy as temperature.

The first law of thermodynamics limits the highest temperature of any atom - that of its shell-1 electrons - to its environmental EME, which means that the temperature of all its other electron shells must decrease with increasing radius. This is the reason the specific heat capacity of elements decreases with increasing atomic number. The upshot of which, means that at any given environmental temperature, metals with a higher atomic number (Z) feel colder to the touch.

5.3 Neutronic Ratio

All elements comprise collections of deuterium (D) and tritium (T) atoms. The neutronic ratio (ψ) of an atom is the ratio of these atoms; T:D = RAM/Z-1

Therefore, the 'neutronic ratio can theoretically be between: $1 < \psi < 2$

However, whilst the theoretical maximum value for ψ is 2, neutron-neutron interaction in atoms with a neutronic ratio (ψ) greater than 1.6 will spontaneously self-destruct. 1.6 is therefore the theoretical upper limit for a neutronic ratio.

Atoms with a neutronic ratio (ψ) greater than 1.5 will continuously eject neutrons as alpha and beta-particles, making them radioactive.

Over time, atoms naturally try to achieve $\psi = 1$, which is their most stable form. They eventually achieve this by ejecting surplus neutrons as alpha and beta-particles or their reversion to proton-electron pairs. The rate at which this occurs is referred to as the 'half-life' of the atom.

To summarise, for any atom:

$\psi = 2$: contains 100% tritium atoms (impossible)

$\psi = 1$: contains 100% deuterium atoms (chemically inactive)

$\psi > 1.6$: will self-destruct (most unstable)

$\psi > 1.5$: will be radioactive (unstable)

$\psi < 1.5$: chemically reactive

Because chemical reactions will not occur between atoms with a neutronic ratio of 1, neutrons are not only responsible for all universal energy, they are also responsible for all chemical reactions; organic and inorganic.

Our universe only exists because of its neutrons.

5.3.1 Nucleic Structure

The lattice structure (ζ) of all matter of any atom is unique to its neutronic ratio and is replicated in elemental matter in both viscous and gaseous states. This uniqueness implies that the current arrangements; close-packed hexagonal, face-centre cubic, tetrahedral, etc. are over simplified. Every lattice structure appears to be unique.

5.3.2 Nucleic Property Table

The following Table lists the nominal neutronic ratio (ψ) and lattice factor (ζ).

Element	Z	ψ	Γ	Noble	ζ
Hydrogen	1	0	0.0715		3.866
Helium	**2**	**1**	**0.0117**	**0**	**9.308**
Lithium	3	1.31	2.823		1.6824
Beryllium	4	1.25	2.2774		1.8068
Boron	5	1.16	1.46		1.7816
Carbon	6	1	0.0161		1.8076
Nitrogen	7	1	0.0086		5.5909
Oxygen	8	1	0.0001		5.902
Fluorine	**9**	**1.11**	**0.9984**	**1**	**6.3248**
Neon	10	1.02	0.1617		8.8979
Sodium	11	1.09	0.8098		2.4543
Magnesium	12	1.03	0.2288		2.6621
Aluminium	13	1.08	0.6795		2.3381
Silicon	14	1.01	0.055		2.3207
Phosphorus	15	1.06	0.5843		3.7295
Sulphur	16	1	0.0366		3.5339
Chlorine	17	1.09	0.7692		4.7768
Argon	**18**	**1.22**	**1.974**	**2**	**6.7282**
Potassium	19	1.06	0.5202		2.6535
Calcium	20	1	0.0351		2.5189
Scandium	21	1.14	1.2668		2.4423
Titanium	22	1.18	1.582		2.576
Vanadium	23	1.21	1.9336		2.744
Chromium	24	1.17	1.4985		3.068
Manganese	25	1.2	0.7777		3.3648
Iron	26	1.15	1.3317		3.0942
Cobalt	27	1.18	1.6444		3.1758
Nickel	28	1.1	0.8657		3.18
Copper	29	1.19	1.7212		3.3402
Zinc	30	1.18	1.6155		4.2666
Gallium	31	1.25	2.2422		3.218
Germanium	32	1.27	2.4216		2.929
Arsenic	33	1.27	2.4332		4.5338

Table 5.1: Noble Gases & Lattice Factors

Element	Z	ψ	Γ	Noble	ζ
Selenium	*34*	*1.32*	*2.9012*		*4.2745*
Bromine	*35*	*1.28*	*2.5467*		*5.5397*
Krypton	***36***	***1.33***	***2.9495***	***3***	***7.4403***
Rubidium	*37*	*1.31*	*2.7895*		*3.3437*
Strontium	*38*	*1.31*	*2.7521*		*3.157*
Yttrium	*39*	*1.28*	*2.5167*		*2.7391*
Zirconium	*40*	*1.28*	*2.5254*		*2.741*
Niobium	*41*	*1.27*	*2.3941*		*2.8527*
Molybdenum	*42*	*1.28*	*2.5586*		*2.9933*
Technetium	*43*	*1.3*	*2.7013*		*3.1672*
Ruthenium	*44*	*1.3*	*2.6734*		*3.2417*
Rhodium	*45*	*1.29*	*2.5811*		*3.3868*
Palladium	*46*	*1.31*	*2.8213*		*3.6129*
Silver	*47*	*1.3*	*2.6556*		*3.8601*
Cadmium	*48*	*1.34*	*3.0771*		*4.9376*
Indium	*49*	*1.34*	*3.089*		*3.6317*
Tin	*50*	*1.37*	*3.3642*		*3.4052*
Antimony	*51*	*1.39*	*3.4853*		*3.871*
Tellurium	*52*	*1.45*	*4.0846*		*4.3626*
Iodine	*53*	*1.39*	*3.5498*		*5.8039*
Xenon	***54***	***1.44***	***3.984***	***4***	***7.3059***
Caesium	*55*	*1.42*	*3.748*		*3.6951*
Barium	*56*	*1.45*	*4.0704*		*3.2524*
Lanthanum	*57*	*1.44*	*3.9324*		*3.0566*
Cerium	*58*	*1.42*	*3.7421*		*3.1312*
Praseodymium	*59*	*1.39*	*3.4944*		*3.1121*
Neodymium	*60*	*1.4*	*3.636*		*3.2778*
Promethium	*61*	*1.38*	*3.3934*		*3.3307*
Samarium	*62*	*1.43*	*3.8265*		*3.928*
Europium	*63*	*1.41*	*3.7091*		*3.7538*
Gadolinium	*64*	*1.46*	*4.1133*		*3.3344*
Terbium	*65*	*1.45*	*4.005*		*3.382*
Dysprosium	*66*	*1.46*	*4.1591*		*3.6671*
Holmium	*67*	*1.46*	*4.1548*		*3.6406*
Erbium	*68*	*1.46*	*4.1372*		*3.6039*
Thulium	*69*	*1.45*	*4.0349*		*4.0733*

Table 5.1 (cont.): Noble Gases & Lattice Factors

Element	Z	ψ	Γ	Noble	ζ
Ytterbium	70	1.47	4.248		4.3955
Lutetium	71	1.46	4.1789		3.5005
Hafnium	72	1.48	4.3113		3.4374
Tantalum	73	1.48	4.3086		3.394
Tungsten	74	1.48	4.3601		3.5064
Rhenium	75	1.48	4.3448		3.5612
Osmium	76	1.5	4.5272		3.7649
Iridium	77	1.5	4.4669		3.9183
Platinum	78	1.5	4.509		4.0626
Gold	79	1.49	4.4392		4.3466
Mercury	80	1.51	4.5664		6.8709
Thallium	81	1.52	4.7093		4.7566
Lead	82	1.53	4.7415		4.4932
Bismuth	83	1.52	4.6605		4.4941
Polonium	84	1.49	4.3909		5.0587
Astatine	85	1.47	4.2339		6.0774
Radon	**86**	**1.56**	**5.0244**	**5**	**7.789**
Francium	87	1.56	5.071		4.9528
Radium	88	1.57	5.1162		4.217
Actinium	89	1.55	4.9579		3.6903
Thorium	90	1.58	5.2038		3.3745
Protactinium	91	1.57	5.1429		3.783
Uranium	92	1.59	5.2854		3.9499

Table 5.1 (cont.): Noble Gases & Lattice Factors

The underlined elements are naturally radioactive.

The **bold** elements are the noble gases (see below).

The **noble gases** appear to be dominated by the number **9**; 2, 9, 18, 36, 54 & 86

i.e. when the factor '$\Gamma = 9.(\psi-1)$' is at or closest to an integer; (**0**) to (**5**). This integer value appears to replace neon as a noble gas with fluorine. The reason for this anomaly may well be that the Periodic Table may not be as definitive as currently believed.

5.4 Ion

Ions are atoms with the same atomic number (Z) but possess an electrical charge owing to unequal proton-electron pairing.

Positive ions (atoms that have lost electrons) possess a positive electrical charge. Negative ions (atoms with additional electrons) possess a negative electrical charge. Negative ions are far less common than positive ions.

Only a few atoms exist naturally as negative ions and they are all non-metals$_n$ except for two, which are semi-metals$_s$:

One additional electron (Group VIIA):

Fluorine (9_n), Chlorine (17_n), Bromine (35_n), Iodine (53_n)

Two additional electrons (Group VIA):

Oxygen (8_n), Sulphur (16_n), Selenium (34_n), Tellurium (52_s)

Four additional electrons (Group IVA):

Carbon (6_n), Silicon (14_s).

Any atom can become a positive ion simply by losing one or more of its electrons from impact with free electrons or a strong external positive electrical charge.

Negatively charged ions are a little more difficult to understand. Additional electrons need to be trapped by the positive charge in protons that do not exist in the nucleus: this shouldn't be possible. However, the nucleic structures of the above non-metal atoms probably have at least one exposed proton that is not protected by a neutron and this means that the additional electro-magnetic electrical charge (e$\square$) generated in it is available to trap passing free electrons

5.5 Isotope

Isotopes are element with the same atomic number (Z) but with varying atomic mass because of unequal proton-neutron pairing. Isotope is an alternative way of saying relative atomic mass (RAM).

An element of iron, with 26 protons (Z=26) and 26 neutrons (N=26) is an isotope of 52. However, in nature, most iron elements have more than 26 neutrons, each of which is allotted its own isotope, e.g. 57, 59, etc.

The following rules apply to isotopes:
1) H^+ can never be fused because they always repel each other
2) All proton-electron pairs within elements are Deuterium or Tritium
3) Theoretically: $1 < \psi < 2$; Practically: $1 < \psi < 1.6$ (see below)

Despite the theoretical maximum value for $\psi = \mathbf{2}$:

*If an element achieves a 'ψ' value of greater than **1.5** it will readily and rapidly eject neutrons as alpha and beta-particles.*

*If an element achieves a 'ψ' value of greater than **1.6** it will split into smaller elements ejecting numerous alpha and beta-particles as it does so.*
***1.6** is the limiting number for elements.*

Over time, elements naturally try to achieve $\psi = 1$, which is their most stable form. They eventually achieve this by ejecting surplus neutrons as alpha and beta-particles. The rate at which this occurs is referred to as the 'half-life' of the element.

However, if all elements in the universe had a neutronic ratio of 1, there would be no chemical reactions (including life).

The neutronic ratio of an element is calculated thus: $\Psi = RAM/Z - 1$

5.6 Fusion

Fusion is the union of proton-electron pairs and/or atoms to create a different element. It is accomplished by applying sufficient force to push the nucleus of one atom inside the electron shells of another.

Atomic fusion can only occur naturally inside cold bodies with sufficient mass to generate the necessary core pressure. This is only possible inside galactic force-centres, the great attractor and the ultimate body, because they are sufficiently massive and they are cold.

When atoms or proton-electron pairs are fused together, a small amount of EME will be released as their electrons rearrange themselves into shells. The energy released, however, will be considerably less than that required to fuse the atoms.

For example, @ 30°C:

> The kinetic energy in a carbon atom = 7.29997E-20 J
> The kinetic energy in an iron atom = 1.26627E-19 J
> Individually, they generate a total of: 1.99626E-19 J
> United (as Germanium), they would generate: 1.34614E-19 J

Releasing: 6.50124E-20 J

> The potential energy in a carbon atom = -1.45999E-19 J
> The potential energy in an iron atom = -2.53253E-19 J
> The energy required to unite these two atoms would be: -3.8598E-19 J

Representing a nett energy input of: -3.20968E-19 J

Therefore, it is necessary to input 3.2E-19 Joules of potential energy in order to release 6.5E-20 Joules of kinetic energy; in other words, in this case you need to input five times as much energy as you're releasing.

Fusion requires the input of energy; it does not generate energy. That is why Hades is cold and 40 years of trials have yet to produce a fusion reactor.

It is also important to understand that the release of energy due to fusion is a single instantaneous event, once accomplished, no more energy will be released.

5.7 Fission

The ejection of neutrons is called radioactive decay, and the time over which it occurs naturally is referred to as its half-life. Neutron energy is released either as:

> *electro-magnetic energy* if the neutron decays into a proton-electron pair but is unable to escape from the atomic nucleus. In this case, the energy is released as EME:
>
> EME = E□/2 = 4.09355561131267E-14 J
>
> and the atom concerned will have become a different element (Z+1 for each trapped proton-electron pair).

Or;

> *kinetic energy* if the neutron's particles are released from the atom. In this case, the proton and the electron will be ejected at velocity:
>
> v = √[2.E□ / (m□+m□)] = 9891525.10667846 m/s (3.3% of light-speed)
>
> the proton of which will impact neighbouring neutrons splitting them into their component parts (a proton and an electron), the protons of which will then split other neutrons; a chain reaction. An ejected proton will not impact a neighbouring proton because of their similar electrical polarity.

Fission can only be released naturally by raising the temperature of a sufficiently massive celestial body's core atoms to neutronic condition through planetary spin. This is the energy released in all bright stars and planets.

The release of neutrons can be initiated artificially by raising the temperature of an atom's two proton-electron pairs in shell-1 to the neutronic temperature (T_n); causing both of them to become neutrons. If this is continued, the atom's neutronic ratio will quickly exceed 1.6, at which time, the atom will eject neutronic particles (protons and electrons), the protons of which will impact and split other neutrons releasing their energies; a chain reaction. Just one atom in any mass will result in the release of much more neutron energy than that required to achieve ψ>1.6 in a single atom. It is this process which occurs in the core of bright stars and planets that have been heated through planetary spin.

The heat we feel from bright stars and planets is fissionable energy released as EME, and the hydrogen atmosphere created at their surface is the residue of total neutronic dissemination; the breaking down of core atoms into proton-electron pairs.

Whilst neutrons are continually ejected by all atoms with a neutronic ratio greater than '1', we refer to those with a neutronic ratio greater than '1.5' as unstable because they represent a danger to life-kind.

Neutron-neutron interaction of a mass of atoms with a neutronic ratio close to or greater than 1.6 will increase the risk of a chain reaction causing it to break apart. The critical mass of such matter is that at which it will become impossible to prevent self-destruction. If its critical mass is achieved quickly enough, the ejected atoms will have nowhere to go so the matter will break apart instantly. This is what occurs in an atom bomb.

Whilst our knowledge today allows us only to exploit such matter using the critical mass of radioactive matter, neutron energy could be released from any matter in a controlled manner if processed correctly. I.e. neutron energy could be safely released from; metals, plastics, earth, rocks, waste, etc., and also from nuclear waste.

Radioactivity can, and should be regarded as a friend, not an enemy.

5.8 Half-Life

Every proton-electron pair inside an element with an atomic number greater than one must have a neutron partner in order for the atom to exist as such.

The term "half-life" of an atom does not mean that an atom will exist twice as long as its half-life.

All elements want to revert to their most stable structure; $\psi = 1$, therefore, over time all excess neutrons ($\psi>1$) will eventually be ejected from all elements. The time over which an atom loses half of its excess neutrons is described as its 'half-life'. This period can range from seconds to billions of years.

For example, if one in every thousand carbon atoms possess two tritium atoms, the time taken to reduce this ratio by half is termed its half-life. It would appear (currently claimed) that the period over which this reduction occurs is fixed and constant, irrespective of conditions.

Whilst the mechanism behind the creation of neutrons is now known, their demise, which is apparently time-driven, has yet to be defined mathematically. However, it is expected that a neutron's decay-rate is linked to both; the internal stress generated by the difference between the curvilinear surface velocities (x25.7) of the electron and its proton due to spin at the time of their union; and the magnetic energy generated due to neutron-neutron interaction within a nucleus.

5.9   How it Works

An atom is a collection of proton-electron pairs (with neutrons attached).
An atomic shell is an orbital radius.
There are two orbits for each orbital radius.

This is why we say each atomic shell holds 2 electrons. Both electrons in any shell orbit their proton partner, so they will have different centres; they are offset. Both electrons in any shell will have the same performance (velocity, energy, force, etc.).

The number of orbital shells in any atom: $N = INT([Z-1]/1.9) +1$

If T is the measured temperature of the atom;

the innermost orbital shell radius: $R_1 = X\square/T$

and the orbital radius of any shell: $R_s = N^o.R_1$

and its temperature: $T_s = X\square/R_s$

You may then use the formulas provided in chapter 4.1 to define the performance of all the proton-electron pairs in the atom. Because electrical charges are dominant, you must define the atom's performance using those in Tables 4.1

5.9.1 Atomic Density

The density of an atom varies with temperature (T). The higher the temperature, the greater the atomic density. It may be calculated as follows:

$\rho = {}^4/_3\pi.R_N{}^3$

For example; the density of iron is 7870 kg/m³

whilst the density of its atom:

@ 273.15 K: $\rho = 0.083888668$ kg/m³

@ 12,412 K: $\rho = 7870$ kg/m³

This 'electron-clouding' allows lone protons (and positive ions) to share spare electron charge capacity in neighbouring atoms at low temperatures, resulting in viscosity and chemical bonding.

Whilst chemical and inter-atomic bonding weaken with increasing temperature as orbital radii reduce, atomic strength and density will rise.

5.9.2 Specific Heat Capacity

The specific heat capacity of an atom is the sum of the kinetic energy in all its proton-electron pairs relative to its mass and its measured temperature (T).

The plot below shows the calculated values for specific heat (ΣKE) for all atoms from Z=4 to 92 compared to the documented values that have been taken from various sources and which are subject to experimental error.

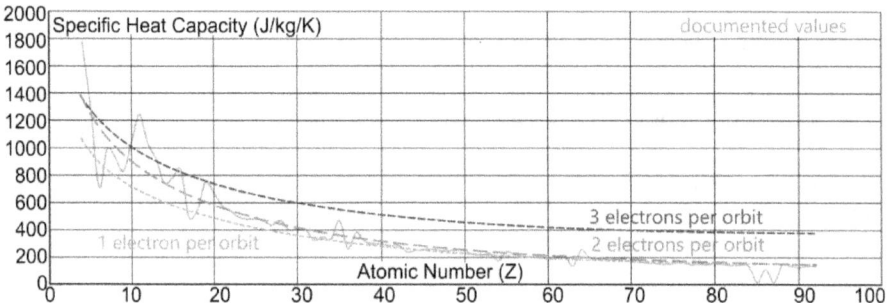

As can be seen, the relationship between the documented values and those calculated for 2 electrons per shell are similar, whereas the calculated values for 1 and 3 electrons per shell are obviously incorrect. This is how we *know* that each shell contains two electrons.

The specific heat capacity of any atom may be calculated thus:

$$SHC = \Sigma KE \, / \, Y.m.T \qquad \{J/kg/K\}$$

where:
ΣKE = the sum of kinetic energies in all the atom's proton-electron pairs
T = the measured temperature of the atom
Y = temperature constant
m = atomic mass (including electrons and neutrons)

6 The Neutron

Neutrons are not particles in their own right. They cannot be picked up (trapped) by accident or design. They are [only] created inside an existing atom that has been exposed to sufficient electro-magnetic radiation (heat). Moreover, they cannot exist outside an atom; if ejected (fission), they will revert to their original components, protons (alpha-particles) and electrons (beta-particles).

They are created by raising the temperature of the proton-electron pairs in an atom's innermost shell to 623316124.717178 K, at which point the electron will be orbiting at the speed of light (c). This is the highest possible temperature that can be generated through natural means. No star or any other body in the universe can exceed this temperature, which therefore represents the maximum possible naturally generated electro-magnetic energy, and is stored in a neutron at the time of its creation.

The heat required to create a neutron can only be generated within a star, and neutrons are the only universal energy storehouse [1].

All the elemental neutrons in all the planets, moons and galactic force-centres were created through fission in stars that existed during previous universal periods. All the elements (large and small) are created through fusion in large, cold bodies: galactic force-centres, the great attractor and the ultimate-body.

During any universal period, all the neutrons in the universe will remain as neutrons unless ejected through natural radioactivity. The percentage of radioactive matter in the earth's constituents is so low that the loss of neutrons through natural radiation is inconsequential.

Every proton-electron pair inside an element with an atomic number greater than one must have a neutron partner in order for the atom to exist as such.

We can refer to the relative quantity of neutrons to protons as an atom's neutronic ratio (ψ), that can only be: $1 < \psi < 1.6$ (chapter 5.3)

The lowest ψ value '1' applies to the most stable elements, that are not chemically reactive.

6.1 Purpose

Neutrons have four primary functions;

> 1) They allow atoms to exist, by shielding proton charges and thereby permitting nucleic protons to sit together inside the same electron shells, and;

> 2) They provide most of the heat and light emitted by the stars (fission), and;

> 3) They will provide the energy for the next 'Big-Bang', and;

> 4) They are nature's energy storehouse for the entire universe.

They are, of course the very essence of universal existence, making the neutron a far more important particle than initially appears. In fact, without the neutron, there would be no elements and no 'Big-Bang' and therefore no universe as we know it!

6.2   Creation

Neutrons are created by raising the heat energy of a proton-electron pair to the neutronic temperature ($\underline{T}\square$). At which time …

… the orbiting electron will have achieved the neutronic radius ($R\square$);

… the orbiting electron will have achieved light-speed (c);

… magnetic field energy will exceed centrifugal energy;

… causing the two particles to unite.

Once united …

… all the energy possessed by the proton-electron pair will be stored;

… the electro-static charges of the proton and the electron will cancel out;

… the magnetic charges (mass) will unite ($m\square = m\square + m\square$).

The [heat] energy required to create a neutron can only be generated naturally within a bright celestial body through planetary spin (friction).

Neutrons are created, and can only exist inside an atom. They are created from the two atomic proton-electron pairs whose electrons are orbiting in the innermost shell.

All the elemental neutrons in all the planets, moons and galactic force-centres were created through planetary spin-friction in bright celestial bodies that existed during previous universal periods.

During any universal period, all the neutrons in the universe will remain as neutrons unless ejected through natural radioactivity. The percentage of radioactive matter in a planet's matter is so low that the loss of neutrons through natural radiation (half-life) is inconsequential.

6.3   The Process

The great attractor and all galactic force-centres together constitute more than 99.99% of all universal matter, and they are very cold; little more than the temperature of outer-space ($\gtrsim$3K). Therefore, when all the matter of the universe re-accretes at the end of a universal period, the heat in all the universe's bright celestial bodies ($\lesssim$0.001%) will be insufficient to raise the temperature of the ultimate body to much above $\gtrsim$4K.

All universal galaxies (force-centres and their satellites) originally comprised similar matter when they were ejected from the ultimate body during the last 'Big-Bang'. They were all cold and lifeless for billions of years. But as galactic satellites (stars) collected more and more sub-satellites (planets) through impact via orbital precession, the satellites began to heat up through planetary spin (friction). When a satellite has collected sufficient sub-satellite mass, it will eventually achieve the neutronic temperature in its core elements.

When an element achieves the neutronic temperature in its two innermost shell (shell-1) proton-electron pairs (all other shell temperatures being progressively colder), these proton-electron pairs will unite to become neutrons. One of two things will occur as a result of this event:

1) Alpha and beta particles, along with their prior neutron partner, will be ejected from the atom (fission).
2) The proton-electron pair will be converted to a neutron and become adopted by a neighbouring deuterium atom, to become a tritium atom.

Once an innermost shell has been lost, all outer shell proton-electron pairs will reduce their orbital radii commensurate with the surrounding heat energy to fill the gap, balancing their electrical charge forces.

6.4 Half-Life?

The term "half-life" of an atom is a bit of a misnomer. It does not mean that an atom will exist twice as long as its half-life.

All elements want to revert to their most stable structure; $\psi = 1$, therefore, over time all excess neutrons ($\psi>1$) will eventually be ejected from all elements. The time over which an atom loses half of its extant excess neutrons is described as its 'half-life'. This period can range from seconds to billions of years.

For example, if one in every thousand carbon atoms possesses two extra neutrons, the time taken to reduce this ratio by half is termed its half-life. It would appear (currently claimed) that the period over which this reduction occurs is fixed and constant, irrespective of conditions.

Whilst the mechanism behind the creation of neutrons is now known (chapter 6.6), their demise, which is apparently time-driven, has yet to be defined mathematically [3]. However, it is expected that a neutron's decay-rate is linked to the internal stress generated by the difference between the curvilinear surface velocities (x27) of the electron and its proton due to spin at the time of their union.

When the universal period concerned is over and all the celestial bodies are reunited as the next ultimate-body, fusion will unite these neutron-rich atoms to create elements such as uranium.

When the ultimate-body next explodes, the planets will now contain matter that was created in the stars during previous universal period(s) and united in the ultimate-body prior to its detonation.

6.5   Proton-Electron Forces

The following Table provides the potential forces either side of the neutronic radius that has been derived from the calculations in Page 66

R {m}	v {m/s}	F_G, F_C & F_c {N}	F_m {N}
2.5812E-15	3.1324E+08	34.62648	37.80183
2.6151E-15	3.1121E+08	33.73676	36.35426
2.6489E-15	3.0921E+08	32.88089	34.97967
2.6827E-15	3.0726E+08	32.05718	33.67351
2.7165E-15	3.0534E+08	31.26405	32.43158
2.7503E-15	3.0346E+08	30.49999	31.24999
2.7841E-15	3.0161E+08	29.76359	30.12510
2.8179E-15	**2.9979E+08**	**29.05355**	**29.05355**
2.8518E-15	2.9801E+08	28.36862	28.03224
2.8856E-15	2.9626E+08	27.70763	27.05823
2.9194E-15	2.9454E+08	27.06947	26.12883
2.9532E-15	2.9285E+08	26.45311	25.24152
2.9870E-15	2.9118E+08	25.85756	24.39392
3.0208E-15	2.8955E+08	25.28190	23.58386
3.0546E-15	2.8794E+08	24.72525	22.80927
3.0885E-15	2.8636E+08	24.18679	22.06824
3.1223E-15	2.8481E+08	23.66572	21.35895
3.1561E-15	2.8328E+08	23.16132	20.67975
3.1899E-15	2.8177E+08	22.67287	20.02903
3.2237E-15	2.8029E+08	22.19971	19.40534
3.2575E-15	2.7883E+08	21.74120	18.80727

Table 4.4: Magnetic Force
R = orbital radius
v = orbital velocity
F_G = gravitational force (according to Newton)
F_C = electrical force (according Coulomb)
F_c = centrifugal force
F_m = magnetic force

The properties in bold type apply to the neutronic condition; $R\square$=2.81793795383896E-15 m

6.6 Neutron energy

If an atom's neutronic ratio exceeds 1.6, its outermost shell neutron(s) will be ejected or remain trapped within the atom, having reverted to a proton-electron pair.

If the particles are ejected, the alpha particles will impact neighbouring neutrons breaking them apart and generating a chain reaction.

If the neutron is trapped and remains in the atom, having reverted to a proton-electron pair, it will release its energy in the form of EME (heat).

The energy a neutron stores at the time of its creation is the electron's kinetic energy, plus the pair's potential energy, plus the spin energies in both particles. Moreover, when a neutron is destroyed, it will take its nucleic proton with it. So, the kinetic energy of an ejected proton must also be included in the sum.

This is the source of energy in atom bombs and nuclear reactors, both of which rely on the critical mass of the heaviest naturally occurring radioactive element (Uranium).

Whilst both processes have served a purpose for the human race to-date, they are; dangerous, uncontrollable, unreliable, inefficient and expensive. But exploited cleverly, fission can provide cheap, clean, safe energy for all. And a simple method for harnessing neutron energy could be made available today.

With little effort, cost and time, we could rid ourselves of all power-stations & wind turbine generators, and almost all batteries & solar panels, all of which are considerably less than 10% efficient, whilst neutrons are more than 231,000,000% efficient. What's more, neutron energy is clean, free and safe to use. Moreover, there are more than enough neutrons to go around; less than one decimetre of the dry surface of the earth's crust is sufficient to fuel our energy needs for an entire universal period.

The neutron is therefore, not only a key feature in the workings of the universe; it should also be the sole source of energy for any intelligent universal life-form.

Important note: All atomic shells contain two electrons. Only the innermost proton-electron pairs convert to neutrons, and they do so simultaneously. Therefore, when neutrons are released, they are usually released in pairs (alpha & beta particles). So twice neutronic energy is released at each instant.

It is here, in all the bright celestial bodies, where neutrons are created and destroyed. And when all the stellar bodies have re-accreted to another ultimate body, at the end of a universal period, they will fuse together naturally due to core pressure. Whilst this final act (fusion) does release a little energy, it is a one-off event, once it has occurred, no more energy will be released.

THE PROTON-ELECTRON PAIR

The electrical charge between a proton and its orbiting electron generates a potential force $(F\square)$ that follows Coulomb's law regardless of orbital radius. The magnetic potential force $(F\square)$, however, varies with orbital radius:

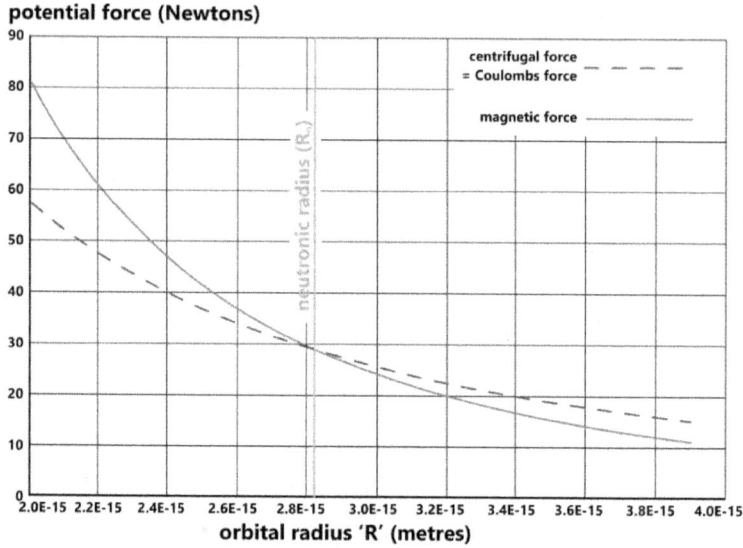

potential force (Newtons)

neutronic radius (R_n)

centrifugal force = Coulombs force -- -- --

magnetic force ————

orbital radius 'R' (metres)

At orbital radii greater than the neutronic radius (R_n), the balancing potential forces of an orbiting electron are as follows:

$$F_G = F\square = F_c > F\square$$

At the neutronic radius (R_n), the balancing potential forces of an orbiting electron are as follows:

$$F_G = F\square = F_c = F\square$$

At orbital radii less than the neutronic radius (R_n), the balancing potential forces of an orbiting electron are as follows:

$$F_G = F\square = F_c < F\square$$

Where the forces are calculated as follows:

Potential: $F_G = -G.m_e.m_p / \varphi.R^2$ $\{m3 / kg.s^2 \cdot kg^2 / m^2 = kg.m/s^2 = N\}$

Electrical: $F\square = -k.e^2 / R^2$ $\{kg.m^3 / C^2.s2 \cdot C^2/m^2 = kg.m/s^2 = N\}$

Centrifugal: $F_c = m_e.v^2 / R$ $\{kg.m^2/s^2 / m = kg.m/s^2 = N\}$

Magnetic: $F\square = \mu.g.e^2 / R$ $\{kg.m/C^2 \cdot m/s^2 \cdot C^2/m = kg.m/s^2 = N\}$

        $= m_e.c^2.R_n^2 / R^3$ $\{kg.m^2/s^2 \cdot m^2/m^3 = kg.m/s^2 = N\}$

and:

$g = -G.m_p / R^2.\varphi$ $\{m^3 / kg.s^2 \cdot kg / m^2 = m/s^2\}$

$\mu = R_n.m_e/e^2 = 1E-07$ $\{kg.m/C^2\}$

where 'R' is the electron's orbital radius.

THE PROTON-ELECTRON PAIR

The energy a neutron stores at the time of its creation includes the kinetic energy in the electron, the potential energy in the proton-electron pair and the spin energies of the two particles at the time of their union. The neutron is not only a key feature in the workings of the universe; it should also be the sole source of energy for any intelligent universal life-form.

The energy stored in all neutrons may be calculated as follows:

$$E_n = |KE| + |PE| + |SE|$$

Kinetic Energy:

$$KE = \tfrac{1}{2}.m_e.c^2$$
$$= \tfrac{1}{2} \times 9.1093897E\text{-}31 \times 299792459$$
$$= 4.09355561131267E\text{-}14 \qquad J$$

Potential Energy:

$$PE = -m_e.c^2$$
$$= -9.1093897E\text{-}31 \times 299792459$$
$$= -8.18711122262534E\text{-}14 \ J$$

Proton Spin Energy:

$$SE_p = E_3 = KE_e + PE$$
$$= 4.09355561131267E\text{-}14 \ J + -8.18711122262534E\text{-}14$$
$$= -4.09355561131267E\text{-}14 \qquad J$$

Electron Spin Energy:

$$SE_e = E_0 - E_1 - E_3$$
$$J_e = \tfrac{2}{5}.m_e.r_e^2 = 7.66586456056651E\text{-}63 \ kg.m^2$$
$$\omega_e = 2\pi/t_n = 1.06387175271756E{+}23 \ \square/s$$
$$E_0 = \tfrac{1}{2}.J_e.\omega_e^2 = 4.33820131944073E\text{-}17 \ J$$
$$E_1 = \delta KE.(r_e/R_n)^2 = 0 \quad J \qquad \{\delta KE = 0\}$$
$$E_3 = 0 \qquad J$$
$$SE_e = 4.33820131944073E\text{-}17 \qquad J$$

Spin Energy:

$$SE = SE_p + SE_e$$
$$SE = |-4.09355561131267E\text{-}14 + 4.33820131944073E\text{-}17|$$
$$= 4.09789381263211E\text{-}14 \qquad J$$

Total Energy:

$$\mathbf{E_n = |KE| + |PE| + |SE| = 1.63785606465701E\text{-}13 \ J}$$

6.6.1 Verification

Apparently, when 'Little Boy' was dropped on Hiroshima ≈ 1.0 kg of its mass was destroyed, releasing; **6.3E+13 Joules** of energy (empirical value).

proton separation:
$$C = 4.R_n^2/(r_p+r_e) = 1.65331294837664E\text{-}14 \text{ m}$$

Proton kinetic energy (according to Coulomb):
$$KE_p = k.e^2/C = 1.3954267683677E\text{-}14 \text{ J}$$

neutronic ratio of U_{235}: $\psi = 1.587270761$

1 kg of which contains:
Neutrons; $N_n = 3.66585231725022E+26$
Protons; $N_p = 2.30953181248165E+26$

According to Newton and Coulomb, 'Little-Boy' released:

$$E_T = N_n.E_n + N_p.KE_p = \mathbf{6.3264167012986E+13 \text{ Joules}} \text{ of energy}$$

6.7 Internal Stress

When the two particles in a proton-electron pair unite they are both spinning according to the following rules:

The surface velocity of the orbiting electron adjacent to its proton:

$$\omega_e = c/R_n = 1.06387175271756E+23 \qquad \square/s$$

$$v_e \approx \omega_e.(R_n\text{-}r_e) = 284361418.5 \qquad m/s$$

The proton's surface velocity:

$$v_p = r_p .\sqrt{[\ 2.|KE+PE| / J_p\ ]}$$

$$= 11062072.34 \qquad m/s$$

Their velocity ratio is therefore: $v_e{:}v_p \approx 25.7$

6.7.1 Hypothesis

It is proposed that at the time of their union, this velocity ratio generates an internal stress between the proton and its orbiting electron, which is responsible for a neutron's decay. It is also assumed that excessive neutron-neutron interaction intensifies this stress, decreasing the period over which the particles can exist as a neutron.

The mechanism responsible for this decay is currently unknown. But it is known that the higher an atom's neutronic ratio (ψ), the more unstable (radioactive) it is, and this means that neutron-neutron interaction causes instability in matter, i.e. neutrons discharge magnetic energy.

7 The Neutron Energy-Cell

As discussed in Chapter 6.6, our current method of fissionable energy exploitation is dangerous, inefficient and expensive, but that is because we only know how to generate it using the critical mass of uranium, whereas, exploited correctly, you can generate fission energy safely from anything, even the rocks in your garden.

I.e. the fuel for a neutron energy generator is free and perfectly safe. You can fuel a car, for example, that will keep going for more than 5000 years from just 1kg of iron, and it is neither flammable nor radioactive.

You can energise your factory, office or house using a neutron energy cell on-site your property or onboard your transport, and the fuel for this energy cell can be extracted from any form of waste or debris.

Despite such an energy source not being in the interests of industrialists and politicians, it must and will be adopted by the human race eventually; after mankind's intelligence conquers its greed?

Neutron energy in the form of heat or electricity can be safely, easily and cheaply exploited, it simply needs the resources to get it started. If you would like to know how to do this, contact the author through CalQlata.

By way of illustration:

the 100g pebble in this photograph possesses 1.37E+12 Joules of neutron [heat] energy; sufficient to fuel:

an average domestic UK household for almost 76 years, or;

an average domestic UK car for almost 100 years.

And it requires no mining. This pebble was found on a beach.

Moreover, nature has provided us with a very simple way to extract it.

8   The Laws of Thermodynamics

The First Law of Thermodynamics: Conservation of energy

Energy can never be lost, it can only be transformed or transferred.

The Second Law of Thermodynamics: Heat will not spontaneously pass from a colder body to hotter body

A high-energy source (hotter body) will spontaneously lose energy to a low-energy source (colder body) but you must add work if you want energy to transfer in the other direction (up-hill so to speak). This law essentially states that it is impossible to create energy from nothing.

This law also claims that energy can, and in fact is, lost by a system to its surroundings but that the reverse cannot happen i.e. an increase in disorder is an inevitable feature of time.

The Third Law of Thermodynamics: The entropy of a substance approaches zero as its temperature approaches zero (absolute)

Entropy is the term used to define disorder. The higher a substance's temperature the more disordered will be its atomic structure and the higher its entropy. E.g. gas has a higher entropy than a solid substance.

9 The Physical Constants

Because everything in the universe is energy, that are defined using only [electrical or magnetic] charge, distance and time, every constant included in this chapter has been defined using *only* these four constants (metric units):

Electricity: Coulomb (C)

Mass: kilogram (kg)

Length: metre (m)

Time: second (s)

Temperature is not a genuine physical variable, but it may be used for the sake of convenience:

Temperature: Kelvin (K)

Notes:
1) Joules and Newtons remain useful, but they are merely compilations of the above.
2) When converting to imperial units ...
... between numerators or denominators: multiply by the conversion factor above,
... across numerators and denominators: divide by the conversion factor above.
*All new constants (unknown until now) are highlighted in **bold** text.*

In the following tables, all of the constants are <u>exact</u> to the number of decimals stated; there are no approximations, tolerances or estimates.

Whilst *mass* is actually *magnetic charge*, I shall continue to refer to it as *mass* in order to prevent confusion.

Whilst *gravity* is actually *magnetism*, I shall continue to refer to it as *gravity* in order to prevent confusion.

9.1 Primary Constants

All universal properties can be defined from only *four* primary constants, *two* ratios and a particle constant (Σ), all of which are listed below:

Symbol	Value	Units
m_e	9.1093897E-31	kg
electron mass (magnetic charge)		
e	1.60217648753E-19	C
electron electrical charge		
R_n	2.81793795383896E-15	m
neutronic radius		
t_n	5.90596121302193E-23	s
neutronic period		
ξ_m	1836.15115053207	
static ratio		
ξ_v	1722.0458764934	
dynamic ratio	(chapter 3.6.15)	
Σ	3E-91 (exact)	m^6
particle constant		
Table 9.1		

Whilst *mass* is actually *magnetic charge* and *gravity* is actually *magnetism*, I shall refer to *mass* and *gravity* in this book in order to minimise confusion.

9.2 Principal Constants

Symbol	Formula	Value	Units
G	$a_o.c^2 / \mathbf{m_u}$	6.67359232004334E-11	$m^3 / s^2.kg$
Newton's gravitational constant			
k	$1/\varepsilon_o$	8.98755184732667E+09	$J.m / C^2$
Coulomb's constant			chapter 3.6.5
k'	$k.RC^2$	2.78024810626745E+32	$m^3 / kg.s^2$
Coulomb's constant (*modified*)			
φ	$G.m_e.m_p / k.e^2$	4.40742111792334E-40	
coupling ratio			
μ_o	$\mathbf{R_n}.m_e/e^2$	1E-07	$kg.m / C^2$
Henry's magnetic constant			
μ	$4\pi.\mu$	1.25663706143592E-06	$kg.m / C^2$
magnetic constant (*spherical*)			
ε_o	$1 / \mu_o.c^2$	1.11265004863082E-10	$C^2 / J.m$
permittivity constant			
ε	$1 / \mu.c^2$	8.85418775855161E-12	$C^2 / J.m$
permittivity constant			
h	$\tfrac{1}{2}.m_e.c.\xi_v . \mathbf{R_n}$	6.62607174469163E-34	$kg.m^2/s$
Planck's constant			
h□	$\tfrac{1}{2}.m_e.c^2 . \mathbf{R_n}$	1.15353857232684E-28	$J.m$
Planck's constant (*modified*)			
ℏ	$h/2\pi$	1.054572071449210E-34	$kg.m^2/s$
Planck's constant (Dirac)			
γ	$(\xi\square / 4\pi)^2$	18778.8808461551	
Rydberg's constant			
a_o	$\mathbf{R_n}.\gamma$	5.2917721067E-11	m
Rydberg's radius			
R_∞	$1 / a_o.\xi_v$	1.09737269561359E+07	/m
Rydberg's wave number			
R_γ	$\tfrac{1}{2}.m_e.c^2 / \gamma$	2.17987197684936E-18	J
Rydberg's electron energy constant			
X	$\underline{T}_n/c^2$	6.9353271647894E-09	$K.s^2/m^2$
heat transfer coefficient (*velocity*)			
X_R	$\underline{T}_n.\mathbf{R_n}$	1.75646616508035E-06	$K.m$
heat transfer coefficient (*radial*)			
Y	$^3\sqrt{[\tfrac{1}{2}.\xi_v]}$	9.51345439232503	
temperature coefficient			
e_n	$m\square.RC$	2.94183820093364E-16	C
proton charge (*neutronic*)			

Table 9.2

Atomic property constants (chapter 8.1; particle properties):

Symbol	Formula	Value	Units
ρ_u	$m_e \cdot \sqrt{[\xi\square/\Sigma]}$	7.1266079635045E+16	kg/m³
ultimate density			
m_u	$\rho_u/1$	7.1266079635045E+16	kg
unit mass of ultimate density			
R_x	$X_R/\underline{T}_x$	8.59854098572228E-07	m
Cold orbital radius			
R_o	$R_n \cdot \xi_v{}^2$	8.3564315638157E-09	m
Planck minimum orbital radius			
R_m	$R_n \cdot \xi_v$	4.85261843362263E-12	m
Planck mean orbital radius			
$R\square$	R_o/κ	8.40016460895157E-11	m
EME orbital radius constant			
v_x	$\sqrt{[\underline{T}_x/X]}$	17162.2425219270	m/s
electron cold velocity			
v_o	c/ξ_v	174090.866621084	m/s
electron minimum orbital velocity (Planck)			
v_m	$c/\sqrt{\xi_v}$	7224342.80705004	m/s
electron mean orbital velocity (Planck)			
c	$2\pi \cdot R_n/t_n$	299792459	m/s
electron neutronic velocity			
$\underline{T}_x$	$X \cdot (c / Y \cdot \xi\square)^2$	2.04274907568265	K
cold temperature			
$\underline{T}_o$	$X \cdot v_o{}^2$	210.193328535837	K
Planck minimum temperature			
$\underline{T}_m$	$X \cdot v_m{}^2$	361962.554671561	K
Planck mean temperature			
$\underline{T}_n$	$X \cdot c^2$	623316124.717179	K
neutronic temperature			
$h\square$	$c \cdot R\square$	8.4479654849081E-07	m²/s
Newton's constant of motion (*electron*)			
κ	$(2\pi)^2 \cdot 2^{4/3}$	99.4793787125405	
EME wavelength (constant)			
K	$t\square^2/R\square^3$	0.15587874533403	s²/m³
constant of proportionality (proton-electron pair)			
e	$\exp(1)$	2.71828182845905	
Natural logarithm			
$\square\square$	$m\square/\rho\square$	7.82336489952175E+46	
$\square\square$	$m\square/\rho\square$	4.26074122343073E+43	
number of particles in a unit mass of ultimate density			
Table 9.3			

Appendices

References, symbols, glossary, etc. used throughout this book along with a summary list of corollaries and hypotheses.

A-1 General

N/A

A-2 References

Most of the references used for the creation of this book are from original work supplied in CalQlata (www.calqlata.com):

The Physical Constants; Keith Dixon-Roche; 978-1-79422-609-8

The Atom; Keith Dixon-Roche; 978-1-08610-029-7

The Life & Times of the Neutron; Keith Dixon-Roche; 978-1-08239-479-9

Orbits; Keith Dixon-Roche; 979-8-33624-956-9

Some additional sources are listed below:

Magnificent Principia; Colin Pask; 978-1-61614-745-7

Seven Brief Lessons on Physics; Carlo Rovelli; 978-0-141-98172-7

Science Data Book; Open University; 0 05 002487 6

Science and Technology Dictionary; Chambers; 0-550-18026-5

A Dictionary of Scientific Units; H G Jerrard & D B McNeill; 0-412-28100-7

It is important to note here that most of the sources here are from work done by pre-20th Century scientists that are universally known and available from sources too numerous to mention here.

A-3 Glossary

Alpha-Particle	An ejected proton
Atom	a collection of proton-electron pairs in which all the proton partners are inside the same innermost shell (Z = 1 to 92)
Atomic Number	The number of protons in an atom (Z)
Atomic Particle	one of the two components in an atom
BCC	body centre cubic lattice structure
Beta-Particle	An ejected electron
Big-Bang	The eruption that occurred when the Ultimate-Body accumulated sufficient 'mass' to compromise the integrity of the innermost neutron
Bright (celestial bodies)	refers to a celestial body that has accumulated sufficient satellite mass to achieve the neutronic temperature in its core matter through planetary spin, thereby creating neutrons, and releasing fissionable energy
Coupling Ratio (φ)	the ratio of the coupling forces due to magnetic and electrical charges: $\varphi = G.m_p.m_e \div k.e^2$
Deuterium	a proton-electron pair with one neutron partner
EME	electro-magnetic energy (it possesses no mass).
Element	a collection of deuterium and tritium atoms (hydrogen to uranium)
Fission	the splitting of neutrons into their component parts; a proton and an electron
Fusion	forcing the nucleus one element inside the innermost electron shells of another
HCP	hexagonal close packed lattice structure
Hydrogen Atom	a proton-electron pair
Neutronic	the conditions that apply to a proton-electron pair at the instant they become a neutron
Neutronic temperature	the highest possible temperature in nature; 623316124.717178 K
Nucleus	the protons and neutrons in the core of an atom
Planetary spin	the relative rotation of a satellite's core and its mantle
Proton-Electron Pair	a proton that hosts an orbiting electron
Quanta	a collective term for all electrons and protons
Tritium	a proton-electron pair with two neutron partners
Ultimate body	A body that contains all the Quanta in the universe ($\approx$2.8E+75 proton-electron pairs)
Ultimate Density	the mass-density of all three atomic particles $\rho = 7.12660796350449E+16$ kg/m³ Nothing in nature has a 'mass'-density greater than this value
Universal period	the time elapsed between 'Big-Bangs'; 31.644 bn-years

All other definitions can be found on the following web page:

http://calqlata.com/help_definitions.html

A-4 Useful Formulas

Equidistant arc-length between 'n' points on the surface of a sphere:

$d = \pi.A / C.n$

where C is the circumference of the sphere

Linear distance across arc-length 'd' (above):

$\ell = 2.R.Sin(\frac{1}{2}.d/R)$

but if you know 'ℓ' and need to find 'n':

$n = \pi / Asin(\frac{1}{2}.\ell/R)$

and if ℓ=R:

$n = \pi / Asin(\frac{1}{2}) = 6$

Lorentz's Equation (magnetic force or field strength):

$F = q.v.B$

Which becomes:

$F = q.g.R.B$

for the laws of orbital motion

Where:

q is the total electrical charge = q1.q2 / me.(q1+q2)

v = relative velocity (electrical circuits)

g = gravitational attraction between m1 & m2

R = radial separation between m1 & m2

$B = \mu\square.e/Rn = Rn.me/e2 . e/Rn = me/e = 1/RC$ kg/C

RC is the relative atomic charge capacity of an electron {C/kg}

$B = 1/RC = 5.685634367312E-12$ kg/C

Inter-atomic force factor (FT):

$\underline{T}\square = \underline{T}\square / \xi\square.Y^2$

$FT = \underline{T}_1/\underline{T}\square$

$\underline{T}_1$ = measured temperature of atom (shell-1 temperature)

A-5 The Heroes

The heroes of this story, to which I offer my gratitude, are listed below

It is not necessary to identify the invaluable contributions made by each of these contributors, they are all widely known and available in almost every scientific publication in circulation today.

Nicolaus Copernicus (Polish) 1473-1543
William Gilbert (English) 1544-1603
Tyco Brahe (Danish) 1546-1601
Galileo Galilei (Italian) 1564-1642
Johannes Kepler (German) 1571-1630
Christiaan Huygens (Dutch) 1629-1695
Isaac Newton (English) 1642-1727
Edmund Halley (English) 1656-1741
Charles-Augustin de Coulomb (French) 1736-1806
Hans Christian Ørsted (Danish) 1777-1851
Michael Faraday (English) 1791-1867
Josef Stefan (Austria) 1815-1863
James Clerk Maxwell (Scottish) 1831-1879
William Crookes (English) 1832-1919
Ludwig Boltzmann (Austria) 1844-1906
Hendrik Lorentz (Dutch) 1853-1928
Jules Henri Poincaré (French) 1854-1912
Johannes Robert Rydberg (Swedish) 1854-1919
Max Karl Ernst Ludwig Planck (German) 1858-1947

The others that were instrumental in the completion of this book are:

My long-suffering wife (Brigitte) sub-editor and critic

My daughter (Eléonore), who initiated this project

Kenneth Pickering friend & editor, who first suggested that I write it

My thanks go out to all the above each of whom have provided a valuable piece of the puzzle without which the final solution would not have been possible, along with my sincere apologies to anybody I have unintentionally omitted.